NEW VIEWPOINTS ON THE ORIGIN OF SQUINT

ISBN 978-94-017-6703-3   ISBN 978-94-017-6770-5 (eBook)
DOI 10.1007/978-94-017-6770-5

Softcover reprint of the hardcover 1st edition 1951

NEW VIEWPOINTS
ON THE ORIGIN OF SQUINT

A CLINICAL AND STATISTICAL STUDY
ON ITS NATURE, CAUSE AND THERAPY

BY

Dr G. B. J. KEINER
OPHTHALMOLOGIST
ZWOLLE (HOLLAND)

SPRINGER-SCIENCE+BUSINESS MEDIA, B.V.
1951

Errata

p. 26,	for	'remarquable'	read	'remarquables'
p. 26,	for	'vieilard'	read	'vieillard'
p. 27,	for	'1000'	read	'900'
p. 27,	for	'mold'	read	'morbid'
p. 35,	for	'forehand'	read	'forehead'
p. 35,	for	'looked'	read	'evoked'
p. 66,	for	'following movements'	read	'following-movements'
p. 95	for	'Degenerative ocular syndrome'	read	'Digitative ocular syndrome'
p. 134,	for	'mental activity-and indeed of visual activity-especially. . .'	read	'mental activity — and indeed of visual activity — especially . . .'
p. 138,	for	'984'	read	'894'
p. 149,	for	'These views'	read	'The views'
p. 150,	for	'2.2%'	read	'2%'
p. 151,	for	'elemintal'	read	'elemental'
p. 175,	for	'Pothological'	read	'Pathological'
p. 211,	for	'984'	read	'894'

CHAPTER I

INTRODUCTION

A survey of impressions gathered during the Strabismus
Symposium held by the Netherlands Ophthalmological
Society in December 1943 leads only to the conclusion
that this Symposium served to show once again how divided are
opinions as to the origin and cause of strabismus. This is a sign
that no single theory is so satisfying or so well-founded that it
can be accepted unreservedly as a basis for further work.

Many important contributions were certainly made, especially
in the fields of reflexology and genetic aspects, but they did not
lead to a new conception surpassing the old theories.

This formed an inducement for a renewed attack on the
problem, with an attempt to gather new facts and data along
new lines of approach in the hope of achieving, if not the construction of a complete theory, at least the establishment of a
sounder foundation for a theory.

In order to become acquainted with the inconsistencies, uncertainties and lacunae in the old theories, we propose to start with
a critical discussion of these. It goes without saying that these
have to be considered and judged according to the period and
the circumstances in which they were evolved.

The problem of squint has occupied the minds of men since
earliest antiquity. It is mentioned even in the papyri of ancient
Egypt and it is probable that an explanation was sought for this
abnormality, which has not only disfigured its victims throughout the history of mankind, but which has also been regarded by
the populace, since the earliest times of which we have any
record, as a stigma of inferiority.

It is a remarkable fact that this instinctive idea of inferiority,
which persisted up to our times, must be admitted in the light
of modern theories of evolution to contain a grain of truth. It

is actually not at all surprising that a function so highly-developed and ontogenetically so young as binocular vision should not reach the same grade of perfection in all individuals at the same time and that it should inevitably bring with it a series of imperfections, the most obvious of which has given its name to the phenomenon.

To regard squinting as a degenerative sign, an involution, however, would seem to be incorrect in the sense of our discuss--ion. It shows rather that even in evolution the path of mankind is an uneven one.

It is understandable that squinting has been regarded through the ages as an abnormality in the position of one or both eyes and that attempts at explanation strove only to account for this abnormal position of the peripheral organ.

Attempts to find an interpretation of this anomalous position of the eyes have been undertaken from various starting points.

The oldest views of all were founded on the idea that seeing as such — the continuous fixation of the light of a lamp hanging to the side, or of an object very close to the eyes, e.g. the tip of the baby's own nose, had caused the trouble to develop. It was supposed that cramp of the ocular muscles was produced and that this had to be counteracted by the presentation. of new objects for fixation, in the form of a lamp placed straight in front of the eyes or pieces of red wool gummed onto the temples 'in order that they shall give these their attention and thereby direct the eyes straight forward' (Paullus van Aegina, 625—691). The use of masks with eccentric holes for the eyes was also recommended by the same writer and this idea was later elaborated further by Bartisch.

These ideas assume that factors such as fixation, convergence and concentration of attention are already present in the infant. The laws which govern the vision of adults were applied uncritically to that of the infant and considered to be valid, an error which has persisted until recent times.

These ideas continued in favour throughout the Middle Ages. Not until long after the end of this period were certain anatomical anomalies of the cornea, lens or retina added to the list of causes of squint (La Hire 1640—1718, Le Cat), while the possibility of a disturbance of muscular equilibrium also received attention. At the same time Le Cat had discovered the existence of amblyopia and used it to explain the non-occurrence of double vision. Petrus Camper (1722—1789), to whom all these ideas were known, draws attention in connection with observations on his eldest son to the influence of gastro-intestinal disturbances on periodical squinting. (De Oculorum Fabrica et Morbis.)

Between him and Dieffenbach (1792—1847), who was the first to perform myotomy on a patient, in 1839, lays a century with comparatively little progress, during which the understanding of the cause and nature of squint failed to keep up with acquisitions in the field of therapy.

In the coming years we shall have to fight out the question of whether the origin of strabismus is to be ascribed to structural or to functional causes.

It is certainly a remarkable fact that in the last 100 years, since Von Graefe laid stress on the importance of mechanical factors in the origin of strabismus, ophthalmologists have continued up to the present day to give these mechanical factors such an important place in their considerations.

This is strange when we consider that the growing knowledge of the anatomy and physiology of the peripheral and the central organ inevitably cast increasing light on the importance of functional factors. As early as the beginning of the 19th. century it was observed (Treviranus, Steinbuch and others) that squinters sometimes developed abnormal correspondence — an observation which was seized upon by both the nativists and the empiricists to provide support for their respective theories.

Mechanical theory: The chief representative of the view that structural factors are the most important for the origin of strabismus was A. von Graefe. He was of the opinion that 'ein Missverhältnis zwischen den durchschnittlichen Muskellängen' was the cause of strabismus concomitans. He also ascribed to accommodation a certain influence on the degree of the abnormality. The final position of the eye was governed, according to this view, by congenital or acquired anomalies of the ligament and muscle systems.

To these factors other workers (Nordenson, Schiötz) added that of the relative positions of the orbital axes.

The nasal displacement of the 'horizontalen Bewegungsstrecke' (horizontal field of gaze) was regarded by Von Graefe as a weighty argument. In the more recent literature Nordlöw allies himself, after an extensive investigation, with Von Graefe's views on this point. In the modern literature of the English-speaking countries the mechanical influences are defended and

Scobee is an ardent advocate of them. After von Graefe we may mention Von Hasner, Schweigger and — in the most recent literature — Nordlöw as the leading protagonists of this view.

Nordlöw studied the displacement of the horizontal field of gaze in 177 'older' strabismus patients; in 66 % of these cases he noted a displacement equal to the primary static angle of strabismus. For another 30.5 % of cases in which this displacemens was 1—13° smaller he states that he is unable to find any explanation. The 3.4 % of cases in which the displacement was larger than the angle of strabismus also remain unexplained. After examining the influence of various other factors (refraction, fusion, correspondence, age of onset etc.) he comes to the conclusion that 'man sich kaum vorstellen kann, dass die Verschiebung auf andere Weise als durch mechanisch wirkende Faktoren entstanden ist', and that these, thus, provide a sufficient explanation of the existence of the angle of strabismus. In his measurements he takes as a basis his observation that the primary static angle of strabismus at 5 M. 'sehr konstant ist und deshalb einen geeigneten Ausgangspunkt für einen Vergleich mit anderen Messungen darstellen dürfte. Er ändert(e) sich während der Messung nur in 2 von 187 Fällen'.

In connection with the determination of the horizontal 'Bewegungsstrecke' it is important to take into consideration the fact that the size of the field of gaze can be ascertained in various ways and with the aid of different stimuli and that the results obtained in these ways will possibly differ considerably among themselves.

Two facts have always commanded special attention in the examination of the field of gaze of squinters: these are:
(a) Limitation of the field of gaze in a temporal direction, from which it was concluded that the abduction was restricted;
(b) extension of the field in a nasal direction, from which it was concluded that the adduction was increased.

Here it is implied that the size of the angle through which the eye can move in the horizontal field of gaze remains unchanged, so that the field is displaced as a whole in a nasal direction. This displacement was observed in both eyes.

For the purpose of determination of the greatest possible movement, use was generally made of an optical stimulus: the following of a moving object or point of light. It is a fact that with this method a restriction of the abduction can be observed in many cases — mostly of rather longer standing — both of unilateral and of alternating strabismus. The fact that this led some investigators to assume that paresis of the rectus externus muscle was present (Snellen), in addition to a mechanical impediment, can be left out of consideration at this point.

It is erroneous, however, to think that the possibility of movement ascertained in the manner described above is the maximal attainable degree of movement.

If, instead of using an optical stimulus, one instructs the patient to turn the eyes as far as possible to the right or to the left, the abduction now proves to be considerably increased, especially with some encouragement.

In addition to the stimuli already mentioned, we can also make use of acoustic and especially labyrinthine stimuli and it is unjustifiable to draw any conclusions as to limitation of function before one has tried these stimuli. In many cases the induction of a compensatory ocular movement by simply turning the head from side to side is sufficient to obviate the error of diagnosis which might have been made with respect to the abduction. If there is still some doubt as to the presence of maximal abduction, there is still a possibility of testing the function with the aid of caloric nystagmus. This test can also be performed without difficulty or danger on babies and very young children, so that it is possible even in these cases to arrive at a well-founded judgement of the function of the lateral rectus muscle. This method of testing for possible abduction disturbances in very young children does not seem to be mentioned in the literature; I take the opportunity of recommending it here.

Nordlöw in his study did not take these different methods of stimulation into consideration. Since it was chiefly a question of the power of movement in the temporal part of the horizontal field of gaze, it would have been possible to check this by means of the vestibular nystagmus. It is now by no means certain that the 'Bewegungsstrecke' determined by him was maximal.

Another objection to the theory of mechanical impediments in general is that it remains to be explained how the eye — which is kept in position by the combined action of all its muscles — can be shifted from its central position by the braking action of certain ligaments or abnormal insertions of one or a few muscles. Moreover, it would be expected that the angle of strabismus would increase than gradually in certain directions of movement.

In addition to this, Von Graefe's ideas are not in accordance with our knowledge of muscle physiology, which does not regard the functioning muscle as having a length of its own but considers the length to be dependent on the function.

Finally, the obviously important hereditary factor in strabismus (at least 28 %) does not tally with the relatively low incidence of anomalies of the muscles and ligaments.

The repeatedly-observed (Roelofs, Piper and many others) disappearance of the squint under deep narcosis is another argument against the mechanical view.

Still more objections to the mechanical theory can be found. Although Nordlöw admits that it is necessary to investigate the three chief factors mentioned in the literature — mechanical factors, refraction and fusion — during the development of the strabismus, in the marshalling of evidence he repeatedly offends, as did his predecessors, against this perfectly correct rule by making use of data on patients from the age of 0 up to 40 years.

His determination of the displacement of the field of gaze was carried out on 177 'older' patients (average age 13 years), in whom, thus, all kinds of secondary compensatory and other reactions had had time to develop, while he used only optical stimuli.

In this way, however, it is not by any means proved that this is the maximal field of gaze; but this is necessary if it is desired to prove that a greater movement of the eyes is impossible for mechanical anatomical reasons. Moreover, the possibility was overlooked that this limitation of movement in the temporal direction might be the result of inhibition or of opposing reflex forces.

The fact that Nordlöw found a fixed angle of strabismus is in itself sufficient evidence that he did not use children in the initial stage for this investigation — and in fact it would have been impossible to do so with this method.

The same objection applies to the work of Scobee, who arrived at his figure of 90 % of mechanical limitations of movement by the study of cases selected on the condition that the abnormality must have developed after the age of six years.

The presence of anatomical abnormalities of the muscle and ligament system has, as such, been observed in many cases. The writer is convinced, however, on the grounds of examination and treatment of many hundreds of children during or shortly after the development of their affection, that these factors cannot be regarded as having any appreciable significance for or effect on the development of strabismus.

The many cases of squint which become cured spontaneously, or as a result of conservative treatment, also point in this direction.

Worth also takes an unfavourable view of the idea that the position of the eye is determined by abnormal insertions, anatomical peculiarities or paresis of the muscles. Of 1,523 cases of convergent squint he found 81 % to have a normal abduction. The disturbance of abduction found in 19 % he ascribes to secondary tissue changes resulting from the prolonged existence of the abnormality. Nordlöw's investigations cannot be regarded as having invalidated these opinions.

The most serious criticism must, however, undoubtedly be directed against the one-sidedness of the mechanical view when one realises that this is concerned with an organ endowed by nature with the means and possibilities par excellence for a maximal degree of mobility, an endowment which seems to be an essential condition for proper fulfilment of the functions of orientation, signalling and alarm-giving for which this organ is responsible.

How did it come about that those seeking the cause of disturbances of position and movement concentrated their attention solely on the anatomical substratum of this movement and neglected the physiological events that precede every movement?

Here we see the unfortunate effect of the continually recurring error of reasoning in which the abnormal position is regarded as the essential of strabismus and the actual cause of the condition is believed to have been discovered once this position has been explained.

It is possible that considerations of this kind led other investigators to look for a central nervous cause for the motility disturbances — which they also assumed to be present.

Muscular paresis, due to aplastic conditions or to obstetrical damage to the centres for the ocular muscles, was suggested as an explanation for the abnormal position.

Apart from the improbably high incidence of developmental disturbances in the primordia of the oculomotor centres which would be required to account for the percentage of squinters among the population as a whole (2—3 %), it is extremely difficult to explain how these originally paretic ocular muscles

manage to function perfectly normally after a few years, a fact
that every clinician has observed times without number.

For the same reason muscular weakness is also difficult to
accept, the more so since physiological investigations have shown
that the normal eye muscle must be regarded as developing, on
contraction, a force nearly a hundred times as large as that
required for a maximal eye movement. (Lancaster) Another fact
which cannot be explained on the basis of these neurogenic
factors is that in cases of congenital or acquired paralysis of the
lateral rectus muscle the eye does not take up an abnormal
position as long as no contracture of the antagonist develops.

Refraction theory. It is greatly to the credit of Donders
that he transferred the problem of strabismus from the mechanical to the biological sphere. He seeks the origin of the disturbance in the coupling of functions of the extrinsic and intrinsic
eye muscles, arising in the binocular fixation of objects at a
short distance. In this way the centre of gravity is shifted to
optomotor factors.

On the grounds of an extensive investigation of visual acuity,
refraction, accommodation amplitude, ocular movements and
angle of strabismus, and also of the age of onset of strabismus.
Donders comes to the conclusion that 'strabismus convergens
usually develops on the basis of hypermetropia'.

As early as 1860 he wrote: 'Since I have examined all strabismus patients for hypermetropia I have been struck by the very
frequent occurrence of the latter. If I am not mistaken, the majority of such cases are actually due to hypermetropia and the
greater prevalence of inward squinting over that in other directions is principally a consequence of this fact. The number of
cases I have been able to examine thoroughly, is, however,
still too small to permit a definite judgement. The question can
only be decided by an extensive statistical study'.

This statistical survey was carried out in 1863 and confirmed
his opinion, the basis of which was the observation that in 77 %
of his 172 cases there was hypermetropia, although as a rule this
was not very pronounced. As a result of the association between
convergence and accommodation these patients are faced, in
Donders' opinion, with the choice between two possibilities:
seeing sharply but not binocular (double) — in which cases the
sight of one eye will as a rule be temporarily or permanently
sacrificed — or seeing binocular but unsharply and with effort.

The increased convergence tendency leads first to periodic squinting which in most cases will develop into a manifest squint. In most cases this process is believed to take place about the 5th. year. He regards inequality of the eyes with respect to visual acuity or refraction, and clouding of the media, as contributory factors. Like Von Graefe he finds increased adduction and restricted abduction in both eyes. This displacement of the horizontal field of gaze he attributes to the increased convergence tendency. Many hypermetropes are satisfied with unsharp images and therefore do not develop strabismus.

Alfred von Graefe, Landolt, Stilling, Maddox and later O. and H. Barkan and Pugh accepted Donders' refraction theory. Landolt assumed further that the strabismus of non-hypermetropic subjects was due to accommodation paresis, while he explained the variable angle of strabismus by the hypothesis that the same accommodation might elicit a variable convergence. Stilling took the view that a combination of refraction factors, in addition to the physiological variability of the relative position of rest of the eyes, was responsible.

The value found by Donders for the percentage occurrence of hypermetropia in sufferers from strabismus convergens was confirmed in later years by several investigators (Lagleyze, 1913). On the whole, Donders' view is still regarded as correct.

But many uncertainties have arisen. In the first place we have the group of myopes, among whom strabismus convergens is not altogether uncommon. Attention was also drawn by Roelofs in 1913 to the ease with which, by means of practice and patience, the association between accommodation and convergence can be broken — in contradiction to Hering's opinion that this association is indissoluble. Nowadays everyone admits that this connection is a loose one. Partly as a consequence of this, the tendency is now to regard refraction more as a pacemaking of contributory factor that can only lead to strabismus when it works in conjunction with other factors.

Another fact difficult to explain was one ascertained by Donders himself, i.e. that in general the degree of hypermetropia found in squinters was not high. Chavasse compared the refraction of 200 normal children with that of 200 children of the same age in the early phase of development of strabismus. All these children were under 5 years of age. The number of non-squinters with hypermetropia of 2—4 dioptres corresponded to 63.5 %, as against 59 % for the squinters. If we take into consideration the ever-present uncertainty as to the exact moment of onset of strabismus, in addition to possible errors in the sciascopic determination of refraction, we are justified in concluding that the two percentages differ only insignificantly.

Statistical data from the writer's own work, dealing with the incidence of hypermetropia among 438 squinting children aged 6 months to 6 years, show 78.9 % to fall into the group 1.5—4 dioptres, with 54.6 % between 2.5 and 4 D. Actually these figures differ but little from those found by Chavasse and it does not seem too venturesome to conclude that there is no significant difference between the refraction of young squinters and that of normal children of the same age group. (see also Chap. IX 2nd section)

Nordlöw calculated the average hypermetropia at the time of onset of strabismus as 3.8 D. He concluded that the refraction factor had 'keine sichere Bedeutung' for the origin of strabismus.

From the statistical data of Nordlöw and of the writer it emerges that in the great majority of cases squinting begins before the second year. In more than half of these cases it is even present in the first year. Obviously the influence of refraction can at this age — on purely anatomical grounds alone — be only very small.

Not without reason, Donders placed the beginning of permanent strabismus after the fifth year. It is, of, course, impossible to conceive the function and the coupling of the accommodation-convergence mechanism otherwise than as grafted onto and supported by a binocular relation. If it should transpire that this had not yet been established at the time when the strabismus commenced, the refraction theory would lose its physiological basis also.

Also important in this connection is Roelofs' investigation which showed that the associative link between convergence and accommodation is built up by means of the fusion; here it is necessary — as will be seen from the writer's work — to distinguish carefully between convergence as a binocular reflex-mechanism and the simultaneous bilateral occurrence of a monocular adduction, i.e. bilateral adduction.

With the gradually increasing anatomical knowledge of the organ of perception and its central nervous connections, together with the consequent improvement in understanding of the physiology of vision, it was inevitable that attention should be turned to a central nervous factor as reason for the abnormal condition. It is certainly no mere coincidence that the extension of knowledge as to the cytological and architectonic structure of the retina and cerebrum — for which we are indebted to Ramon y Cajal and his school — was accompanied by the raising of an increasing number of voices in favour of a search for the origin of strabismus in central nervous disturbances.

Fusion theory: Claud Worth (1869—1936) came forward as spokesman of this trend and the publication of his work heralded a new turn in ophthalmological thought, while in the same period Ramon y Cajal presented his theory of the physiological utility and 'raison d'être' of the chiasma.

On the basis of extensive clinical research and statistical data Worth came to the conclusion that fusion disturbances were capable of accounting for strabismus. He entirely rejected mechanical causes and also considered that refraction could be excluded as a causal factor. The fusion faculty, the mental process of image-formation which culminates in the final fusion of the two sensations to a single perception at the highest level, is the great force that creates and realises the duality-in-unity of binocular vision, while at the same time it determines the correct position of the eyes in space.

Worth points out in the first place that the fusion faculty takes its place beside and above the motor co-ordination of the eyes — which is already partially developed at birth. The inborn motor co-ordination suffices only to maintain the normal position of the eyes relative to each other during the first few months of life. Later on the fusion faculty takes over this function and where it is normally developed there will be no possibility for any agency whatever to divert the eyes from their correct position.

Worth assumes that the fusion faculty is already present at the age of six months, that it has increased considerably by the first birthday and that it is not completely developed until the age of 6 years. The fusion faculty may be congenitally absent or may develop late, incompletely or not at all. 'In the lastmentioned case, any cause tending to disturb the unstable equilibrium, which is here dependent solely upon the motor co-ordination, will cause permanent strabismus'. His conclusion runs as follows: 'The essential cause of squint is a defect of the fusion faculty'. This defect he believes to be present to a greater or smaller degree in every case of convergent unilateral squint. For strabismus alternans he assumes, on the grounds of amblyoscopic investigations, that there is a total congenital absence of the fusion faculty.

As further support for his views Worth points out that in many cases of unilateral squint the fusion faculty can be improved by training and he also refers to his examination of 157 younger brothers and sisters of squinting children, in which he found that 106 of these children had a well-developed fusion faculty while only one of them subsequently developed periodical strabismus. Of a second group of 37 children with a doubtful fusion faculty, eight started to squint at a later date; of the remaining 14, whose fusion faculty was very poor, eight developed a permanent squint.

Worth did not define what he meant by fusion faculty. In the few pages which he dedicates to its development he places it on the same line as 'the desire for binocular vision'. Elsewhere he speaks of 'the instinctive tendency to blend the images formed in the two eyes'. In present-day literature too, the concept is as a rule inadequately defined and the different interpretations placed on it by different authors are a frequent source of confusion. International standardization of various terms used in ophthalmology is badly needed.

What we are to understand here by fusion faculty is as follows:
(a) The fixation of both eyes on the object that attracts attention (the fusion movement);
(b) The fusion of the two sensations into a single perception (fusion in the stricter sense or synchisis).

It would appear that Worth did not have a clear mental picture of these two components from which the fusion faculty is built up. He seems to have thought chiefly of synchisis.

From the beginning of this century Worth's ideas on the rôle of the fusion faculty have dominated ophthalmological literature, although in later years they were subjected to attack from various sides. As recently as during the war years they were opposed by Waardenburg, in this country, on genetic grounds. Nevertheless, both in the English-speaking countries and in Germany there are still many who consider the fusion disturbance to be highly significant as a causal factor in the origin of strabismus. (Cords: Kurzes Handbuch f. Ophthalmologie)

The proofs which Worth brings forward for his theory are not over-convincing. The transformation of strabismus alternans into strabismus unilateralis, in addition to genealogical research and study of twins, shows that there is no essential clinical or genetic difference between these two groups and that the sharp distinction which Worth draws between them with respect to fusion faculty is not justified.

In addition to this, various investigators (Waardenburg, Clausen) have reported cases in which it was possible, after operative treatment in cases of strabismus alternans, to ascertain the existence of binocular vision and fusion faculty. Worth also admits that the fusion faculty takes several years to develop and that prolonged covering of an eye, or a congenital defect of function, impedes its natural development. How are we now to explain the 'congenital' squint which is present in more than 18 % of cases — in which any exercise of the fusion faculty is a priori excluded on the grounds of the faulty position — and with the total of 54 % in which the child already squints before or about its first birthday! All this holds both for the alternately and for the unilaterally squinting child. When Worth concludes that these children squint because their fusion faculty is deficient he is to a certain extent confusing cause with effect. These children do not squint because they have no fusion faculty: they have no fusion faculty because they squint. As regards the examination of brothers and sisters of squinting children, it should be pointed out that sufficient evidence has accumulated in the course of years to show that even a primary disturbance of fusion — if there is such a thing — does not as such necessarily lead to squinting, and that all kinds of predisposing factors (unequal acuity of vision, infectious diseases etc.) play a certain part in this respect. For an accurate judgement of the different groups it is necessary to know these factors, and also the ages of the children. The only information that now remains to be gathered from these figures relates to the markedly-familial nature of the affection (10.8 % of the brothers and sisters subsequently developed strabismus).

Worth and many others believe the fusion faculty to be hereditary in the same way as — to quote the words of Scobee — brown eyes are hereditary. Others, however, conceive and define it as a reflex or a combination of reflexes, whereby they seem to be thinking chiefly of the fusion movement. Chavasse, who defends this standpoint, expresses himself as follows: 'We need no longer vainly gesticulate before the fireless altar of defect of the fusion faculty any more than we need to be content to regard lameness (with which strabismus has so much in common) as a defect of the walking faculty'.

In his study of fusion movement Roelofs pointed out 'that the fusion movement only takes place if the light stimuli falling upon two disparate retinal areas penetrate, with intensities that are sufficient and do not differ too greatly, to that part of the cerebral cortex where they give rise to conscious perceptions'.

The fusion faculty as a whole, thus, is to be regarded as a psycho-optic reflex. Roelofs also points out that where the fusion movement is absent strabismus will be present, either because the normal fusion movement was insufficient on account of excessive heterophoria or because there is a disturbance somewhere in the very complicated mechanism of the fusion movement.

An important causal factor of strabismus may, thus, lie not in a disturbed synchisis as believed by Worth but in a disturbed fusion movement. In this way Worth's 'defect of the fusion faculty' may arise in renewed glory from the ashes of Chavasse's 'fireless altar'.

In subsequent years the conflict between the protagonists of the refraction and the fusion theory drew attention to various important factors which are observed in connection with strabismus but which these theories neither explained satisfactorily nor — in the opinion of many — took sufficiently into account. The chief of these, which were also unmistakable interconnected, were undoubtedly diplopia, suppression of one image and amblyopia. These factors were considered to be of such importance that many investigators came gradually to the conviction that they were, either alone or in combination, largely responsible for the development of strabismus.

The amblyopia which is so common in strabismus pointed the way to a more thorough study of diplopia as expression of a disturbed retinal correspondence. This led Van der Hoeve to ascribe an important rôle in the causation of many cases of strabismus to 'diplopiaphobia', the fear of double vision, and to the consequent development of purposive squint. Facilitated as it was by the peripheral situation of the double image, the suppression of this image was promoted by the purposive squint. This suppression of the double image, whether active or passive, has continued up to the present day to prevail in ophthalmological thought as an important factor in the production of squint. For amblyopia also, the controversy of effect versus cause is still unresolved.

Primary, congenital amblyopia and amblyopia ex anopsia still have their protagonists and antagonists, although the odds seem to be more in favour of the ex anopsia party. The fact that it is possible to cure amblyopia ex anopsia, to induce it or to cause it to alternate, simply by covering one eye in young children with a tendency to squint, leads Van der Hoeve to conclude, rightly, that there is a definite connection between these phenomena, although he is of the opinion that further investigation, especially along statistical lines, is highly desirable for the provision of greater certainty, more particularly in view of the fact that he estimates the number of amblyopic eyes in the Netherlands at more than 50,000. It is well to reflect for a moment on this figure and to realise that this amblyopia due to disuse of the eye could have been prevented or cured, with an enormous saving of happiness, working capacity and expense.

Another controversial point was — and for some apparently still is — the question of wether the affection is a unilateral or a bilateral one, while far more important is the question of whether the affection is really not peripheral but central in nature and whether the position of the eyes, which has for many centuries been regarded as the affection itself, does not deserve to be regarded rather as one of its symptoms.

It is true that Stilling and Arlt had already expressed the opinion, in the 19th. century, that the affection was bilateral in character (strabismus alternans in itself points in this direction), but the belief in a unilateral affection persisted for a long time, partly as a result of the general acceptance of mechanical peripheral factors as the causative agents. The idea that we are concerned here with a bilateral disturbance of senso-motor reactions based upon peripheral and central causes has, however, gradually gained ground.

Theory of Van der Hoeve. A study of the anatomical and topographical relations in the orbit had led Van der Hoeve to the conviction that the achievement and maintenance of the physiological and relative position of rest of the eyes required both good binocular vision and a mechanism of continuous regulation and correction of the innervation of the ocular muscles. Here the possibility of monocular correction must also be present, since the resistance encountered by the eye upon movement in the orbit is not necessarily the same for both eyes, so that an equal innervation impulse for both eyes might produce an unequal effect.

Starting from the premise that every deviation from the correct position of the lines of sight will lead to double vision, and that this double vision is experienced as something unpleasant — although to an individually-varying degree — Van der Hoeve sought in diplopiaphobia, the fear of seeing double, the stimulus which sets going the muscular mechanism of escape from this double vision. He regards diplopiaphobia as a psychological factor, a psycho-reflex if this term be preferred, in the same way as the fusion tendency and the inhibition which he believes to act to a certain extent as aids to the diplopiaphobia.

Van der Hoeve has worked out his ideas to a very ingeniously constructed theory, which he expounded once more in the supplement to the Symposium on Strabismus. He postulates that the problem of strabismus is the problem of the position of the eyes with respect to each other. This may be an ideal or a squinting position. The position of the eyes is governed by various factors, such as static factors, dynamic factors, influence of accommodation on convergence etc., each of which can cause a form of squint.

The position of the eyes is determined by the algebraic sum of the angles of strabismus caused by each of these various forms of squint separately. This sum may be zero or have a measurable value. Where the different forms of squint do not cancel out, the position of the eyes is a squinting position. If the position is a squinting one, the image of the surroundings does not fall on corresponding areas of the two retinae. This leads to double vision — by which we are to understand the entire complex of perceptions that may by caused in this way, including the double image of the object fixed (in the fixing eye falling on the macula and in the other eye so far distant from the macula as corresponds to the angle of squint); the double image of the object that is represented at the macula of the non-fixing eye and in the fixing eye so far from the macula as corresponds to the angle of squint, in addition to the fact that the macula of the two eyes are occupied by different objects, which may lead to competition as to which object is to receive the most attention. This double vision is experienced — but with large individual differences — as something unpleasant (diplopiaphobia), whereupon the subject endeavours to avoid seeing double, to escape from the squint.

This can take place in various ways, viz:

(1) escape towards the ideal position, leading to more or less complete fusion;

(2) complete or partial suppression (inhibition) of the image belonging to one eye, with or without the acquisition of a new form of binocular vision by the development of a pseudo-macula, whereby the true macula with its connected regions is usually excluded as such.

Van der Hoeve now examines the results with respect to the position of the eyes which can be achieved by the escape from double vision in either of the ways defined under (1) and (2). If the subject escapes from double vision by achieving fusion, this fusion movement will bring about a new form of squint in which the angle is equal to the algebraic sum of all other forms of squint, but with the opposite sign. The whole is then a latent squint.

If the escape is achieved by suppressing the image of one of the eyes in situ, the result will be a manifest squint which the diplopiaphobia makes into a concomitant squint with an angle of strabismus equal to the algebraic sum of the angles of strabismus of the component factors, among which escape squint is not included.

If the escape is achieved by suppression of the image in one eye at another place at which suppression is easier, the result will also be a manifest squint but now which an angle of strabismus in which purposive squint is represented among the component factors.

Finally, if the escape is effected by exclusion of the true macula and development of a pseudo-macula, the result will be concomitant squint, among the component factors of which purposive squint is absent or present according to whether the pseudo-macula is developed at the same or another place.

As final result we thus arrive at the following different types of relative position of the two eyes:

I. The ideal position:

(a) Without any form of squint or with two or more forms of squint which cancel out, and do not include purposive squint (orthophoria).

(b) With two or more forms of squint which cancel out and which include purposive squint (heterophoria).

II. The squinting position:

(a) due to one or more forms of squint which do not cancel out and which do not include purposive squint;

(b) due to one or more forms of squint which do not cancel out and which do include purposive squint, either with or without the formation of a pseudo-macula (manifest squint).

None of these factors gives a squint with a constant angle. The changes are governed by the changes in the double vision. The diplopiaphobia is then the stimulus which brings the eyes back to the correct position by means of purposive squint. In this way orthophoria, heterophoria and concomitant squint are continually guarded by diplopiaphobia with purposive squint.

Although Van der Hoeve's theory offered an explanation for many things which had previously remained unexplained, it also left several questions unanswered. It is not altogether easy to form a conception of the manner in which a fear of seeing double can form or elicit the conditioning stimulus necessary for monocular correction of an undesirable position of the eyes: in other words, can a psychological process form a link in a chain of physiological processes? If not, what is the physiological process running parallel to this psychological one? Are we to place Van der Hoeve's desire to avoid double vision on the same line as Worth's desire for binocular vision? Are the disagreeable sensations to which diplopiaphobia gives rise really caused by the double vision?

If we confine ourselves to clinical facts, the fact that congenital or early acquired abducens paralysis does not lead to purposive squint requires some explanation. Is is really necessary, in view of the possibility of inhibition even of the macular image of the deviating eye, to imagine that the paramacular image of this eye first has to undergo displacement, by means of purposive squint, for that purpose?

Roelofs sought an explanation along experimental lines for some of the above-mentioned problems. He came to the conclusion that the disagreeable sensations accompanying double vision were the result of abnormal muscle innervations called into being for the purpose of correction. Therefore, they could not possibly be the cause of these innervations.

He does not believe seeing double as such to be unpleasant.

It is only unpleasant to see two objects when one knows or believes that there is only one.

Both these views are supported by the fact that clinical experience, including my own, shows that the unpleasant sensations are felt before the double images appear, and that they disappear when the double images become manifest.

Finally we come to another general question. If we assume, as is now generally done, that strabismus begins between the ages of 2 and 4, what idea are we to form of the vision of these children before the beginning of their affection? Are we to assume that they were in possession of binocular vision with a 'retinal correspondence' in the course of development or already established, according to their age?

Without this it is difficult to conceive how double images can trouble them. It is obvious, however, that the ‚congenital' cases of squint and those developing in the first year cannot in this way be accounted for by diplopiaphobia and purposive squint. The sharpness of images is still poor at this period and it is not yet possible to speak of a genuine retinal correspondence'.

Donders had already attempted to explain the adduction tendency that is present in squint in another way. Priestly-Smith, Parinaud and, still later, Deloge and Stutterheim, followed him in this respect and drew attention to the possible rôle of convergence in the production of strabismus.

Innervation theory of Duane. About 1934 Duane assembled these views in the formulation of his innervation theory, which was based on the premise that the action of abnormal, non-adequate innervation impulses on the convergence mechanism, expressed in the form of an augmented convergence or divergence stimulus, was responsible for the conditions arising in squint.

In addition to these fluctuations of tone of the convergence centre (the existence of a divergence centre is denied by many investigators), Duane also makes room in his theory for mechanical and other factors interfering with binocular vision. The source of these abnormal impulses, however, remains obscure, so that this theory also does no more than shift the difficulties without solving them. It has, however, many convinced adherents especially in America.

It was inevitable that the great advances in the physiology of muscle and nerve which were achieved by Sherrington and Pavlov and their schools should have their influence on the thought of those investigators who were not satisfied with current ideas and wished to test the existing theories on the origin of strabismus in the light of newer points of view.

The therapeutic results so far achieved had been, apart from the aesthetic effect of surgical methods, so disappointing that the need for further investigation was obvious.

Sherrington's work on the reflexes and that of Pavlov on the part played by conditioned and other reflexes in the production of many vital manifestations and performances, including those connected with the senses, opened up entirely new avenues of thought for ophthalmological physiology.

It gradually became clear that the boundary which Pavlov had originally drawn between conditional and unconditional reflexes was only of theoretical interest and that in actual practice these two forms merge into each other. So wide was the field that they ultimately came to embrace, that scientists began to doubt whether there was sufficient justification for retaining the concept of 'voluntary' action or expression at all.

These were not entirely new ideas in the physiology of ocular movements. As early as 1889 Wernicke — and soon after him Roux — had drawn attention to the markedly reflex character of these movements. In the Netherlands, Reddingius wrote in the same period: 'Beim Sehen ist nur willkürlich: das lenken der Aufmerksamkeit; im übrigen kommen alle Augenbewegungen als Reflexe zustande'. Roux was but little less positive when he remarked: 'eine willkürliche Augenbewegung ist eine zusammengestellte Reflexbewegung, vergesellschaftet, eingeleitet von der Vorstellung der Augenbewegung und der Illusion: Freie Wahl.'

Wirth points out that in the slow following of moving objects there is a summation of stimuli starting from central and paracentral points on the retina, resulting in an extra impulse. Zeeman and Roelofs, like Hering before them, also draw attention in their work published in 1929 to the reflex character of ocular movements. The very complex combination of convergence, accommodation, fixation and fusion movements is still regarded as a reflex process, although many workers believe that an important place must be left open for the idea of 'attention'. In his study of latent nystagmus, Van der Hoeve in 1918 placed the great influence of optical and labyrinthine stimuli on the achievement of fusion and, on the other hand, on the braking of the tendency to nystagmus, in their correct light. It is remarkable that, even with this view of the character of the position of the eyes and of ocular movements, the great stimulus provided by Pavlov's admirable research was necessary before the reflex nature of the eye movements in strabismus was recognized as such, and before conditional stimuli originating from the peripheral organ were accorded the place that is theirs by right. Shortly before the outbreak of World War II, Chavasse had realised that the achievement of binocular vision required the co-ordination of two monocular reflex systems, the co-adapation and coupling, in sensory and motor respects, of a paired organ to give a single functional unit at a higher level. It was obvious that this must involve interference, summation and inhibition of stimuli, these changing with and being dependent on the demands of the situation. Only slowly, however, did the conviction gain ground that such a complex mechanism, in order to be effective, could only operate along reflex paths, and that any disturbance in the perceptor or effector organ would find expression as a disturbance of binocular vision and would be manifested externally as a disturbance of co-ordination.

Reflex theory of Zeeman. Our gratitude is due to Zeeman for the presentation, as the fruit of many years of thought and study, of his reflex theory, in which the various impulses are assigned to their proper rôles in the establishment of normal and disturbed binocular vision in accordance with their mode of origin, their nature and the path which they fellow. This he did as one of the contributors to the Strabismus Symposium — that admirable testimony to the moral fibre of the Dutch ophthalmologists — under the crushing burden of the Nazi occupation in what was for them probably the most disastrous year of the war.

In his inimitable way Zeeman shows us how, during the growth of the developing organism, the ever-changing relationships create conditions and possibilities for the formation of new reflexes, how at birth and during further growth new links are continually added to the existing reflex chain for reasons of efficacy and necessity; along what routes and pathways the conditional stimuli reach their goal and their intended effect and thereby themselves may give rise to new reactions which will subsequently become law.

He sketches the commandingly important potentialities which

must be ascribed to light as a conditioning stimulus; the difficulties which may arise in looking and seeing, as a result of the multifarious and often frankly conflicting stimuli which reach the sensomotor centre of the eye from all sides and the manner in which — at one moment by means of summation and at another by total or partial inhibition — the final effect can be attained only by way of harmonious co-operation.

Starting from the assumption that the optomotor impulses will make use of existing reflex paths for the centrifugal portion of their journey, he distinguishes three substrata:

(1) A binocular reflex grafted onto the vestibular reflex path;
(2) a monocular reflex grafted onto the proprioceptive reflex path;
(3) a convergence reflex grafted onto and adjusted to the monocular reflexes and thus also ultimately based on the proprioceptive reflex paths.

He shows us the eye in its connections and correlations with other reflex centres, with the stimuli which it thus receives from all parts of the body, and finally he defines the demands which must be fulfilled for the achievement of binocular vision which requires composition and co-operation of the stimuli arriving simultaneously from both eyes, in order that these may be unified at a higher level. The problem can be solved in three different ways, as follows:

(a) harmonious satisfaction resulting from equilibrated reactions which fulfil both demands;
(b) opposed and irreconcilable impulses which prevail in turn, resulting in alternating vision;
(c) suppression — inhibition of one of the two demands.

Zeeman believes the choice between these solutions to be dependent upon various internal and external factors. By analogy with the observations of Pavlov, it is possible that the failure to achieve normal binocular development may have its cause in unfavourable conditions at that time of life at which accommodation and retinal differentiation come into being and undergo adjustment.

All this, however, does not explain the squinting position of the eyes but accounts only for their simultaneous, alternating or unilateral use, as the case may be.

As keystone for his hypothesis Zeeman — like al those who

preceded him — is obliged to call in heterophoria as a conditioning factor.

In explanation of the concept of heterophoria Zeeman points out that the reactions evoked by the retinal impulses act upon an apparatus that has already been exposed to the influence of all kinds of stimuli coming from other sense organs (proprioceptors, labyrinths) since long before birth. These innervations may — in the same way as congenital or acquired malformations or defects, in addition to internal or neurogenic factors — have the result that a normal harmonious development of the optical reflexes is *a priori* excluded. Here we have, thus, an outline of the factors which Zeeman regards as constituting the primary cause of disturbances in the development of the optomotor reflexes.

Under the leadership of the reflexes evoked by the various impulses mentioned above, different degrees of co-ordination of the two eyes are achieved. In many normal individuals a very high degree of co-ordination, which we call orthophoria, is reached; in a very great number of individuals, however, the result is heterophoria which expresses itself in an anomalous position of the eye when it is covered and excluded from vision.

The cause of this heterophoria may lie in the refraction, in the stimuli coming from other sense organs, in internal and neurogenic factors and sometimes also in congenital or acquired morphological conditions or defects (mechanical resistances).

The continued absence of corrective optomotor reflexes will have the result that the heterophoria may develop into a permanent squint of one type or another.

Zeeman thus regards existing heterophoria as a condition for the establishment of a manifest squint. In a cursory reference to the cause of squint he agrees with Nordlöw in attributing great importance to mechanical factors.

Zeeman's ideas are worthy of mention at this point in discussion of the theories of squint — although he himself does not intend them as such — not only because they come so near to providing an explanation of strabismus but still more because they have done so much to enlarge and enrich our ideas as to the nature, the building up and the fate of the reflex factors involved therein.

Zeeman regards squint as the 'expression of a change in the normal course of events with respect to the reflexes involved in vision'.

His aim in the Symposium was not to seek the cause but — through a search for and investigation of the normal reflexes and ascertainment of the changes present in these in squint — to provide a physiological basis for therapeutic measures and thereby to increase their efficacy.

If we may venture a few comments on Zeeman's opinions, these will be concerned chiefly with that part of his address which dealt with the position of the eye.

In accordance with his view of squint as 'fundamentally a disturbance of binocular vision', Zeeman is obliged to seek the beginning of squint at an early age. Since, however, it follows from the degree of maturity of the perceiving and of the central organ, and from the associated development and acquisition of function of the optomotor reflexes, that binocular vision cannot come into play as a rule before the 2nd. 6 months of life, it is not possible to hold a failure of binocular vision responsible for those cases in which the squint appears shortly after birth or at any rate at a very early age (one can safely say in the 1st year). The number of such cases is by no means negligible; the present writer has shown statistically that it must be computed at not less than 54 % of the total number. One is inclined to doubt whether Zeeman, in the formulation of his definition, accorded to these cases the attention which they deserve. In these cases developing at such an early age the definition 'disturbance of binocular vision' does not seem altogether fortunate, since binocular vision is not yet completely developed in normal babies at this age.

As regards heterophoria, Roelofs has pointed out that the great importance which is generally attached to this in connection with the development of strabismus is not in accordance with the simple dominant hereditary character of strabismus, in view of the numerous causes which are held to be capable of giving rise to heterophoria.

We are left with a number of questions: what is the primary cause of a disturbed development of reflexes; is there an individual or group predisposition to abnormal development and can we give a name to these predisposing and/or heriditary factors? All these must remain unanswered for the present. We cannot explain why it should happen to one individual and not to another.

In this way squint remains for us what it was before: *a*

secondary affection developing under the influence and action of various factors. The reason for this is our insufficient understanding of the normal physiological conditions in general and a lack of knowledge of the pathological physiology of squint in particular.

When we realize, further, that nothing is known as to the morbid anatomy of the clinical condition and that our knowledge of the postnatal growth and the process (extending over many years) of maturation of the brain and the eye of the child is insufficient, in short that our anatomical and physiological foundations are still lacking, we cannot be surprised that no satisfactory explanation of strabismus has yet been found and that opinions differ as to the nature and importance of various factors connected with the syndrome.

Primary versus secondary amblyopia and amblyopia ex anopsia are still highly controversial ideas, while other factors which are found upon careful clinical examination and are more or less generally accepted, such as fixed angle of squint, nasal displacement with unchanged extent of the horizontal fields of gaze, central scotoma in amblyopia, frequency of anisometropia, effect of refraction on the development of squint, age of onset and many other points, are still open to dispute.

At the beginning of this introduction we remarked that there was a great need for renewed investigation. We are not alone in this opinion. Some very recent utterances in the ophthalmological literature show, more clearly than any words of ours can do, how widespread is the realization of this need. These voices also give expression to the disappointment felt with the poor progress made in the last half-century. In connection with the therapy, S. V. Abraham writes:

„The lack of understanding as to cause and effect, and the lack of clarity in the interpretation of response to various therapeutic procedures are reflected in the multiplicity of the theories, methods of treatment, and definition of terms.

More accurate evaluation of the conditions leading to and following the onset of strabismus will permit emphasis on the factors which should more properly be considered primary or exciting factors and those which should be considered secondary factors. The effect that it may be of more importance to treat the secondary factors and sequelae, in cases which have existed for any considerable time should not blind us to an appreciation of cause and effect. In only this way can a truly attitude be developed in outlining the prophylactic and specific therapeutic measures necessary in the treatment of strabismus."

In a survey of literature on the oculorotary muscles covering the last 6 years, R. G. Scobee writes:

„If one reads a bibliography of all of the literature on this subject since 1943, the list is indeed impressive, not only in length but in ponderous titles. One gains the impression that we are sweeping forward in an advance toward new horizons.

Thoughtful and careful examination of the individual articles however, reveals a startling lack of progress and a good deal of preoccupation with minutiae. This is not to say that the many published papers are not worth-while but simply that only a comperatively small number represent real advances — advances that push back the barriers of our knowledge toward ever distant horizons." He continues with a quotation from Adler:

„It has been pointed out, that convergent heterotropia is not a single entity produced by one etiologic factor, but is an abnormal position of the visual axis due to a number of different causes. Therefore, no single theory as to the cause of esotropia satisfactorily explains all of the cases one sees.

Esotropia may be either partly or entirely accommodative, paralytic or undetermined. In the „undetermined" group, the deviation often disappears under general anesthesia and hence would seem to be due to an abnormal convergence innervation, although the sources of abnormal convergence tone are not well known. A close inspection of the various reflexes influencing the two eyes might yield valuable information about the sources of "undetermined" cases of both esophoria and esotropia, according to the writer".

Finally we have the opinion of Onfray:

„Tandis que depuis quelques années les progrès de nos connaissances et de notre thérapeutique ont été si remarquable, en d'autres domaines, la question des mouvements oculaires et des strabismes n'a pas depuis une demie-siècle, fait de pas décisif. Cela tient certainement à l'imprecision des donneés anatomiques et à l'absence total d'une anatomie pathologique des strabismes non paralityques.

„En attendant le chercheur patient qui fera des coups en série du mesencéphale d'un enfant strabique et ou mieux d'un vieilard atteint depuis l'enfance d'un strabisme convergent alternant, nous devons nous contenter des traveaux de tous ceux qui s'attachent à préciser les connaissances physiologiques et physiopathologiques de l'appareil oculo-moteur."

Encouraged by these and similar expressions of opinion, the writer ventures to present in this book his ideas and opinions on the cause, occurrence and nature of strabismus convergens concomitans, opinions which are based on five years' experience in the study and treatment of nearly 1000 cases and which represent original and entirely new points of view.

A fundamental principle of the investigation was that the affection must be studied exclusively during or immediately after its production, i.e. in very young children, before secondary abnormal compensation reactions have had the chance to obscure or modify the original picture. Indispensable for a correct judgement and interpretation of the observed facts was a preliminary study of the anatomical and physiological relationships and possibilities in the normal baby and the growing child. Innumerable lacunae in this knowledge required to be filled by original investigation. All explanations had to be consistent with the age and to be supported by a knowledge of the anatomical and physiological conditions existing at that moment. In this way we avoided the ever-recurring error of simply equating the vision of the newborn and the young infant to that of the adult and, mutatis mutandis, applying to the infant the observations made on adults.

The first necessity was then to ascertain the actual age at which squint begins. In contradistinction to the generally accepted view, this proved to be sometimes shortly after birth and sometimes at a quite early age, a discovery which made it possible to draw valuable conclusions.

Starting from the ocular movements observed in normal infants during and immediately after birth and in their first weeks, in addition to observations of ocular movements in the mold condition of 'papilla grisea', we were able to make a list of factors which may be held responsible for movement or rest and also to determine the position of the eyes in space. It is scarcely necessary to point out that everything that happens at this period of life takes place exclusively in a reflex manner, by means of conditioned or other reflexes. Anyone who thinks that a baby a few weeks or months old is going to trouble its head about double images has an erroneous idea of its physiological possibilities and of the nature of growth.

By analogy with what has been observed and described in connection with children suffering from 'papillea grisea', a search was made for myelination disorders in squinting babies, and evidence was obtained that such existed.

The result of this is a disturbance of the physiological equilibrium of the motor stimuli transmitted from the retina, leading to an abnormal position of one or both eyes.

By means of electroretinography it could be shown that the cause of this disturbance was almost certainly not in the retina itself.

In this way we find that an eye which is born with small imperfections in its optomotor paths suffers functional defeat from the 'defects of its good qualities', i.e. the impossibility of establishing the normal binocular linkage which is essential for the achievement and maintenance of binocular vision.

The results of the writer's statistical, clinical and physiological investigations proved to be incompatible in certain respects with the current views. This was enough to induce him — albeit with a certain diffidence — to attempt the construction of a new hypothesis as to the cause and nature of *convergent* strabismus.

He is fully conscious of the many lacunae and imperfections, which means that his work too can only be called an attempt, while he is acutely aware of the lack of those results of anatomical and pathological investigations which would be required to provide its keystone. The peculiarly extensive nature of the subject has compelled him to confine his investigations, for the present, to those which are reported here.

'History repeats itself' and this applies also to the history of science in its upward spiral. More than half a century ago, Parinaud in his address to the 'Société Française d'Ophthalmologie' on the treatment of squint, found himself obliged to request his colleagues to forget — even if only for a few hours — their traditionally acquired knowledge. At this moment the present writer finds himself in a similar position and ventures to address a similar request to his readers.

CHAPTER II

EXAMINATION OF SQUINTING CHILDREN

'IT is the task of the clinician, by seeking the normal reflexes in the patients in question, to ascertain the absence or disturbance of these reflexes and — in so far as the relevant anatomical and physiological factors are known — to draw conclusions herefrom as to the site, nature and causation of the altered mode of reaction, in order to gain from and in these data a knowledge of the point of attack, the desired change and the possibilities of effective treatment.'

In these words of Zeeman's lie the starting point and the aim of my investigations. It was disappointing to find that even now our knowledge of the anatomy and physiology of a great part of our organ of sight, and in particular that of the newborn and the young child, is full of lacunae. This is true especially of the postnatal development of the brain. We ophthalmologists are to blame for the fact that it is not yet possible to speak of any soundly-based anatomical and physiological knowledge of the postnatal development of the eye and its connections to the central nervous system. But such knowledge is indispensable for a correct understanding and interpretation of clinical observations. The present state of the strabismus problem serves to show us that — as in the time of Boerhaave — all theoretical considerations that are without such foundations are doomed to sterility.

Where knowledge of this kind is lacking or inadequate, painstaking clinical observation — in addition to experimental investigation — has always been found of great value as a source of fresh knowledge. In many cases it is this clinical observation that serves for the formulation of problems for experimental study and that subsequently submits the conclusions drawn from experimental research to the acid test of comparison with the facts.

Clinical observation also serves to open the eyes of the practitioner daily — in his interpretation of observed phenomena or search for causes — to the gaps in his knowledge, thus stimulating and fertilizing his striving for a broadening and deepening of his knowledge.

The writer feels that the dividing line between theoretical and applied science which has become deeper and wider, chiefly for social reasons, has caused a great loss to science by impairing the necessary co-operation and reciprocity between these two branches.

There are many problems which can only be solved with the aid of data from actual practice. Only in this way is it possible for a single observer to collect, screen and arrange a large body of data, while the precision and significance of the investigation are increased by the large number of observations and the long period of time which it covers. Examples are genetic and refraction problems (myopia; emmetropization) and many others. The strabismus problem also belongs to this class.

In addition to the medical duty imposed upon him, I believe it to be the scientific duty of the clinician to keep proper records of his cases and, if possible, to arrange and analyse them. Where the latter is impossible for him, he should hand over his data to one of the Universities. Should a general regulation of this kind prove impracticable, it would at least be possible to provide the Universities with facilities for establishing contact with a group of specialists, in order by means of team-work and regular contact and exchange of information to serve the interests of scientific progress and at the same time to prevent the loss of a great store of valuable observations.

In the clinical examination of squinting children the following fixed schema was followed in all cases. The points were noted in the following order:

1) anamnesis;
2) position of the eyes;
3) ocular movements and optomotor reactions;
4) pupil reactions;
5) refraction;
6) fundus;
7) visual acuity;
8) optical localization;
9) general condition and behaviour;

10) other disturbances, if any.

re 1). In the anamnesis were noted: disorders of pregnancy; duration of pregnancy; birth; time of appearance of the squint; what illnesses the child had had, whether he/she had been normal or late in walking, talking and cleanliness; left- or right-handedness and any evidence of a hereditary nature of the squint.

re 2). The position of the eyes was noted both in the light and in the dark-room. The type and angle of strabismus were ascertained.

re 3). The movements of each eye were studied separately and for both eyes together. In the light this was done with an object attractive for the child; in the dark-room it was done with the directed pencil of rays of the 'Oculus' ophthalmoscope. The compensatory ocular movements were studied. Finally, true or apparent convergence movements were sought.

re 4). In cases of confirmed or suspected blindness, investigation of the pupil width and pupil reaction is particularly useful.

re 5). The refraction was determined after 4 days with 3 applications per day of drops of 0.5 % atropine sulphate. After skiascopic readings, check tests with lenses followed. The refraction thus found was taken as the correct value. Astigmatism was determined with Javal's instrument.

re 6). After the refraction determinations the fundus was examined with the pupil dilated. Particular attention was paid to the macular region.

re 7). The visual acuity was determined for each eye separately by means of the various pictures on Burghardt's chart, which is particularly suitable for this purpose. For very young children we used white or coloured objects (beads) of various sizes, held in the examiner's hand or placed on the ground.

re 8). Optical localization was determined only for the amblyopic eye, also with Burghardt's chart.

re 9). Intelligence, concentration, irritability and nervous condition were also noted, in addition to the general physical condition.

re 10). Attention was paid to motor and functional disturbances as well as anatomical anomalies.

The anamnesis. This is extremely important. The data furnished should be checked where possible by interrogation of other members of the family, by examination of baby-photos or portraits of the child's first years and by examination of relatives. The sooner the child is examined after the beginning of the affection, the more reliable will be the information obtained. The time when the child was still an infant-in-arms, its first steps and its first words are all landmarks in the mother's memory and will serve as a check on her answers to repeated interrogations, and where her recollections are not clear they will help her to assign the beginning of the squint to its proper time. The older the child is when seen for the first time, the more essential does this checking become. A remarkable fact is that the grandmother is often more observant than the mother where squinting is concerned.

The position of the eyes. On inspection of this, a manifest deviation is immediately obvious, whereas some patience is needed before a periodical or slight, incipient deviation can be detected in ordinary daylight. I have found from experience that blinking with one or both eyes may be the first objective sign of a very inconspicuous, slight, periodical squint in a child with an apparently labile binocular adjustment; the child blinks or winks from time to time to get rid of the double images. This is seen rather often in cases where there is familial occurrence of strabismus. The true cause is usually revealed only by the examination of optomotor reactions. For this reason it is advisable to include these reflexes as a matter of course in the routine examination of children.

The first thing which strikes the examiner when the child looks at him is that, in young children with a unilateral squint, the angle of squint is highly variable and may range from $0-30^\circ$ within a few minutes. This is most noticeable where the squint is still of fairly recent data and the amblyopia is not too pronounced. With patient observation, however, it can be seen in all young squinting children. In the dark-room the position is usually a different one and the angle of strabismus is on the average smaller. With older children some influence of refraction factors (accommodation) may also be considered in this connection, and with younger children perhaps a lowering of tone of the gazing centre. Upon subsequent examination of the ocular movements the varying magnitude of the angle of strabismus once more attracts attention. It is not necessary to measure this accurately; it is sufficient to estimate it when the child looks one in the face. It will be found that this angle averages about 15° in strabismus unilateralis, while in strabismus alternans it is often larger at the beginning.

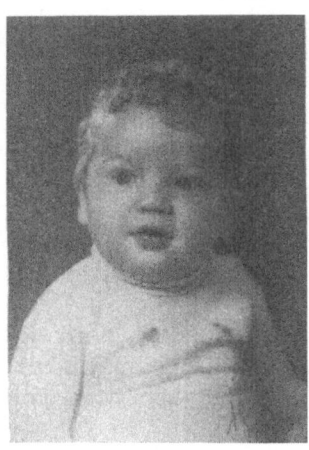

Typical case of squint not observed by the parents in a child aged 9 months.

With older children and adults whose squint has been neglected for many years, the angle of strabismus seems to be more or less fixed, owing either to secondary changes in the tissues or to an abnormal correspondence. On closer examination, however, this is also found to be not entirely constant.

Where the two clinical types of strabismus seem to merge into each other, the best guide to a classification of doubtful cases is the visual acuity. If this is not the same on both sides, while no objective cause of the dissimilarity is found, it should be concluded that amblyopia is present and the case be placed in the unilateral group. This classification is exclusively of clinical significance, as has been shown by genetic studies (Waardenburg).

In addition to deviations in the horizontal plane, those in the vertical plane are also noted. The percentage of manifest deviations in this direction among my cases is small, amounting to only 9 cases out of 568 in children under 6 years of age (1.58 %). Abraham also, in his examination of 1.054 cases of convergent squint, found only a few vertical deviations.

The data on this point in the literature, however, vary widely. Ringland Anderson found such deviations in more than half (55 %) of his cases of strabismus convergens concomitans. He does not, however, state the ages or the exact method of examination used.

It is worthy of note that he also seeks the cause of the vertical deviations exclusively in mechanical factors and pareses, along the lines which we have already discussed in connection with the mechanical theory of squint. Opinions on this point are indeed practically unanimous (see also White and others).

The present writer is convinced, however, that the influence of abnormal optomotor reactions must certainly be taken into consideration for these cases, just as for the horizontal deviations.

The alternating, exclusively vertical deviation (alternating hyperphoria), which Chavasse regards as physiological, falls outside the scope of this investigation. Moreover, as will be obvious from the foregoing remark about the causes of vertical deviations, the writer is not inclined to ally himself with Chavasse on this point.

Ocular movements and optomotor reactions. Optical, acoustic and labyrinthine stimuli can be used for the examination of eye movements. We start with optical stimuli in daylight, with the aid of a moving object. It is important to use a small object and one that arouses the child's interest. A finger is not the most suitable object for this purpose. The writer used a little soapstone monkey measuring 1½ cm.

Starting with the child looking straight ahead, once his attention has been attracted to the object, the maximal ocular movement is determined, first binocularly and then with each eye covered in turn.

When the maximal ab- and adduction positions in the horizontal field have been thus ascertained, the movement in the vertical field is tested in the same way.

In an examination of this kind it will be noted that a limitation of abduction for one or both eyes, found in the binocular test, often proves to be absent in the monocular tests.

If it appears from the results of these tests that there is limitation of movement in one direction or the other, it is advisable to verify the conclusion with the aid of acoustic or labyrinthine stimuli. One should then try, by snapping the fingers, to induce the child to carry out a maximal abduction (it is usually the abduction with which we are concerned).

If this acoustic stimulus does not succeed either, even after a certain amount of encouragement, in eliciting a maximal abduction position, one should make a further attempt — by a simple turning of the head to the right and to the left — to achieve this by way of a compensatory eye movement. If the movement still remains submaximal, we can finally call in the aid of caloric nystagmus to provide information as to the functional possibilities of the abductor. Not until one has tried by all these means, but without success, to evoke a normal function is one justified in concluding that a disturbance of motility is present.

This part of the examination in daylight is concluded by

persuading the child once more to fix his eyes on the small object, which is first held about 40 cm. from his eyes and is then moved in a straight line in the median plane to a few cm. from his eyes. Any adduction or other movement of each eye elicited in this way is noted. An impression of the accommodation range or of the presence of an accommodation-adduction coupling can be obtained in the following way: one hand is held in the median plane and supported against the child's nose and forehand, thus blocking the sight of the object for one eye, while the object is brought close to the other eye.

We have now entered the realm of the optomotor reflexes and we continue the examination of these in the dark-room, where the optomotor reactions evoked by a peripheral light stimulus to both eyes and to each eye separately are studied. The stimulus used is the directed pencil of rays of the 'Oculus' ophthalmoscope, with which a very satisfactory isolation of the stimulus can be achieved.

The pencil of rays is now directed alternately, from a very short distance, first on the temporal and then on the nasal half of the retina of each eye, the effect on each eye separately and on both together being noted.

We found in our squinting child-patients a striking difference between the reactions which followed stimulation of the temporal half of the retina and those observed after a stimulus striking the nasal half. The reactions were chiefly monocular and characterized by a predominant adduction tendency.

Finally we test the reactions to a peripheral light stimulus which strikes the homonymous retinal halves simultaneously and those to a bitemporal stimulation of the retinae.

The results of the examination of ocular movements and optomotor reflexes looked in this way have greatly improved our understanding of the nature of squint and have proved to be of great importance in the search for the true cause of this affection. We propose, therefore, to analyse and discuss them in detail in a subsequent chapter (IV), to which we refer the reader for further particulars.

Pupil reaction. The careful testing and repeated re-testing of this is one of the most important methods for the differentiation of the various degrees of blindness which may be found in newborns. Not only the fact but also the nature and the speed of the pupil movements must be noted, in addition to the pupil width.

Refraction. This has always occupied a very important position in the literature on squint, especially since the time that Donders drew attention to the possible causal connection with hypermetropia. Although it soon became evident that this factor, notwithstanding its significance for the production of a periodical or permanent squint in certain predisposed cases, was liable to be over-rated, the influence of Donders' theory is still with us and even in recent literature one repeatedly comes across publications in which this refraction anomaly is regarded as one of the chief causes of squint and the strabismus cases are classified according to the magnitude of this anomaly or the degree to which the two eyes differ with respect to refraction.

The current American view sees the refraction as an activating factor which is capable of influencing the state of tonicity of the convergence mechanism in a positive or negative sense, leading to a change in the anatomical position of rest of the eyes (Duane, Chavasse, Abraham, Peter). For the purpose of further explanation of the part played by ametropia in the causation of squint, Chavasse actually classifies the cases into 10 refraction groups. (Scobee: 'The oculorotatory muscles'), while Abraham in a recent study attempts a new classification by dividing his 1.054 patients with strabismus convergens 'from newborns to greybeards' into an isometropic and ananiso metropic group, thus characterizing the influence of refraction as a causal factor for each of these groups.

In speaking of refraction in connection with strabismus cases we refer always to the total refraction, i.e. that found after the introduction of atropine drops on several consecutive days. One or two applications of drops of homatropine in the course of a ½ hour, as frequently practised as a matter of routine in the consulting room, is entirely insufficient for this purpose. In my cases drops of 0.5 % atropine sulphate solution were introduced into both eyes three times a day for four days in succession. I have found from experience that in this way the ciliary muscle is to all intents and purposes put out of action. On the morning of the day of examination the drops are given an extra time, to be on the safe side.

In the examination which now follows the position of the eyes is first ascertained once again, first in daylight and then in the dark-room. In a few cases the abnormality will now be found to have disappeared. This may be a sign of so-called accommodative squint, but it is by no means necessarily so. In a few other cases the deviation is found to be smaller in darkness than in daylight.

Here again, where the accommodation has been eliminated, we still see continual variations in the angle of strabismus, both in daylight and darkness. In addition to this, in very young children the accommodation cannot — for anatomical reasons — have played a very important part in these variations.

It is repeatedly observed that after several days' atropinization even the somewhat older child seems not to be troubled in the least by the elimination of his accommodation. He is not in any way handicapped in his play but continues to build towers and play with small objects and has no difficulty whatever in finding small objects which have fallen on the floor.

The optomotor reactions are now tested again in the dark-room. Then the fundus is examined and finally the skiascopic refraction is determined; here it is important to take care that the amblyopic eye fixes correctly. This however, as noted by Wald and Burian in the course of their investigation, does not as a rule give any difficulty in darkness. Where the patient's age permits, the astigmatism of the cornea is now measured with Javal's apparatus. The values found are checked with lenses and the total refraction is also determined in this way to an accuracy of $\frac{1}{2}$ D. The resulting figure is accepted as the correct one.

With children under 2 years of age one is generally obliged to content oneself with the objectively determined values, against which there is no theoretical objection. The child is now provided with spectacles such as to give the fullest possible correction. This is not usually done with children younger than 1 year. If however, it is desired to make a baby of this age wear spectacles, this is quite possible — as proved by experience — even at the age of 8 months.

For very young children it is advisable to fix the two ear-pieces together with a piece of elastic running round the back of the head. This prevents the child from taking off the spectacles — a thing of which we actually hear very few complaints.

Of course the baby tries at first to take off its spectacles — but it does the same with its bonnet or shoes, and it becomes accustomed to the spectacles just as soon as to those useful objects. If desired one may prescribe plastic lenses, but I have never found this necessary.

The first thing that strikes me on examining the results of the refraction measurements is the rare occurrence of refraction errors greater than 6 D. This confirms the view of Donders as to the influence of the degree of hypermetropia in the causation of squint, in despite of the criticism to which this view has been subjected in recent years. Among my cases in children under 6 years of age I have found only a few such cases. The most frequently found errors were between 1.5 and 4.0 D; these sometimes increased a few dioptres in the next few years and in some cases they subsequently decreased again during the school years.

Here we touch on an important point. Too little attention is paid — also in the everyday practice of refraction testing with children — to the process of emmetropization. In the world literature too, this fact, first studied by Straub, seems to be insufficiently recognized.

The next point which emerges is that in our series of very young patients there is only a very small percentage (2.3 %) with a difference in refraction between the two eyes; this is in marked disagreement with the figures in the literature (Lagleyze; Abraham).

Abraham found 70 % of isometropia against 30 % of anisometropia, but his patients ranged 'from newborns to greybeards'. Lagleyze found anisometropia in 44 % of his cases. It is a common opinion that in strabismus the amblyopic eye has a greater error of refraction than the good eye. In my cases, however, I found anything but confirmation for this opinion. On the contrary, with a few exceptions all young squinting children have the same refraction in both eyes.

Again and again children are brought to me with spectacles prescribed for them elsewhere, which would appear to give grounds for the assumption that anisometropia is present. In almost all cases I have found after 4 days' atropinization that the refraction of both eyes is the same. Honesty compels me to remark that the error usually turned out to be greater than would have been expected from the spectacles prescribed.

I believe that the notable difference between the figures given in the literature and the percentage of anisometropia found among my cases can be accounted for by the youth of my patients. When older patients with strabismus unilateralis and amblyopia are examined one does indeed often find a difference in refraction between the two eyes, the amblyopic eye then as a rule having the greatest error. It is possible that we are here concerned with the failure of a normal process of emmetropization to occur.

Finally, I found only a small percentage of astigmatic cases, while the maximal refraction difference between the two meridians was only 3 D.

In this connection it is advisable to keep in mind the fact that in young children the refraction factor cannot — for anatomical and physiological reasons — be of more than subordinate importance, whether as a predisposing or other factor. This in contradistinction to some older patients, in whom the accommodation — as becomes evident as soon as the spectacles are taken off — has for years served as a contributory factor in determining the magnitude of the angle of strabismus.

In actual practice these so-called accommodative cases are indeed not found in very young squinters. It seems to be a question chiefly of academic interest whether such cases really occur at all and whether the forms of squint in very young children, in which these reactions are supposed to be present are not in fact due to other reactions (e.g. fusion disturbances, variations of tone etc.).

It is of great importance that the share of the refraction as a causal factor in the development of squint be known as accurately as possible. A survey of the results of refraction measurements may be very useful in the assessment of this share. The following tables and a graph show the results of examination of 437 squinting children between the ages of 6 months and 6 years.

Table I

Dioptres	Hypermetropia 0—7	Myopia —	Astigmatism 1—3	Anisometropia —	Total
No. of cases	369	5	53	10	437
%age	84.4	1.2	12.1	2.3	100

Survey of the refraction of 437 cases of strabismus convergens in children aged 6 months to 6 years.

Table II

HYPERMETROPIA

Refraction in D	0.5—1	1.5—2	2.5—3	3.5—4	4.5—5	5.5—6	6.5—7	Total
No. of cases	0	90	110	91	45	30	3	369
%age	0	24.4	29.8	24.6	12.2	8.2	0.8	100

Survey of the refraction of the 369 cases of hypermetropia (See table 1)

From the table I it is seen that 84.4 % of these children were hypermetropes as against only 1.2 % myopes; that astigmatism to a maximum of 3 D occurred in 12.1 % of the cases and that only 2.3 % showed a (usually small) difference in refraction between the two eyes. From table II, that in 78.8 % of the cases the refraction ranged from 1.5—4 D and in only 0.8 % was there a hypermetropia of more than 6 D.

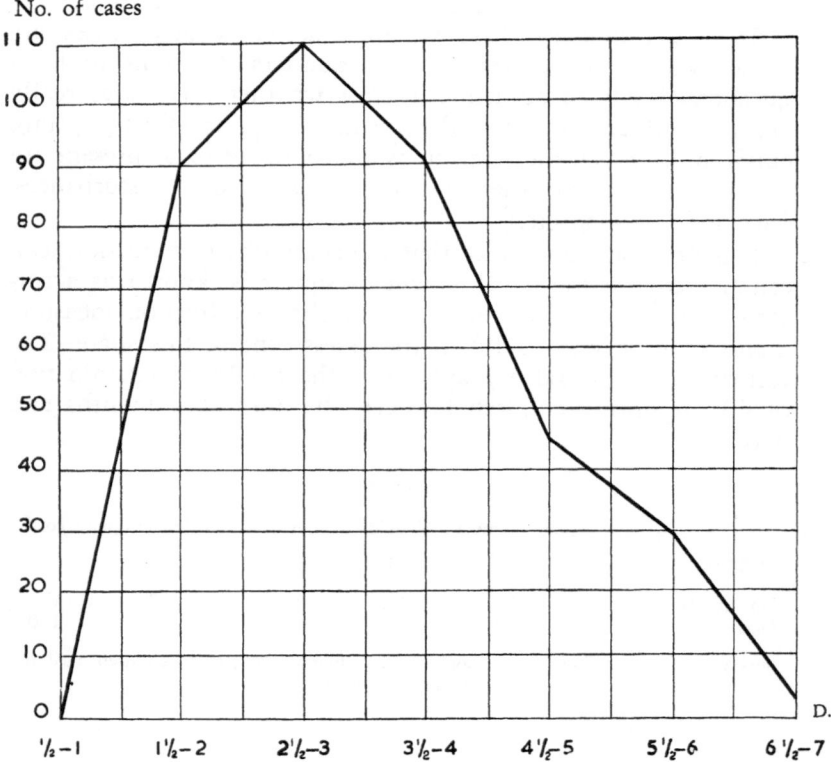

Refraction curve for 369 cases of hypermetropia in squinting children aged 6 months to 6 years.

This curve, corresponding to 369 cases of strabismus convergens, shows that the overwhelming majority (78.8 %) of these cases fall into the lower refraction group (1.5—4 D.). Cords reaches the opposite conclusion on the grounds of the statistics of Worth and Lagleyze, in which the highest frequency is observed in the group of 4—6 D. The cases studied by both these authors, however, came from all age groups and are therefore not comparable with mine.

The values given above permit the following conclusions:

1. The refraction percentages found differ only slightly from those accepted as averages for normal children of the same age (See Chavasse, Chap. I).

2. Nearly 79 % of the squinting children belong to the lower refraction groups (up to 4 D). The incidence of greater errors is extremely low (less than 1 %).

3. Anisometropia is very rare in young squinting children (less than 2.5 %).

From points 1, 2 and 3 we may deduce:

4. that refraction cannot be considered to play more than a very small part in the causation of squint.

For the group of 'congenital' cases or those developing in the first year of life, any influence of the refraction can a priori be regarded as excluded.

Visual acuity: The first examination of this was done with the aid of Burghardt's figures [1]). These seemed to me very suitable for my patients, among whom were practically no city children. The pictures used must be such as to attract the child's attention; they must be simple and must be interesting to him. The exactitude of the data obtained will in many cases be enhanced by the use of figures to illustrate a fairy-tale (e.g. a cat and a boot for 'Puss in Boots') in order to stimulate interest. Care should be taken to avoid disapproving remarks from the parents, which often make the child nervous, and to help and encourage the little patient as much as possible. The result is often directly proportional to the amount of trouble taken and it will be found — as so often — that experience and patience bring wisdom. One should not forget to give the child his reward.

Repeated checking of the findings is necessary. From the age of 2 years onwards it is usually possible in this way to make a fairly accurate judgement of the visual acuity of a child of

[1]) A test chart with simple pictures.

42

normal intelligence. Proper covering of the eye which is not
being examined is essential. For younger children one is obliged
to have recourse to coloured beads of different sizes, small toy
figures or sometimes sweets — (fruit drops, chocolates or sugar
animals). These are held in the hand in front of the child while
one or the other eye is covered in turn. From the child's
behaviour it is possible to judge the state of his vision and to
ascertain whether there is strabismus unilaterialis with amblyopia or strabismus alternans. In case of doubt one should let
the child take hold of the toy or induce him to reach out for
it, while covering each eye in turn in the most imperceptible
manner possible.

Sometimes a combination or comparison of the results of
the first with those of a subsequent examination will be necessary before one can get a correct idea of the visual acuity. Difficult
though this often is, one must be prepared to spend the
necessary amount of time on the examination of the young
squinting child.

After the refraction has been determined the examination
of the visual acuity is repeated after correction with lenses.
Essential for this purpose is a simple spectacle frame made to
fit a child's head and permitting easy and rapid changing of
the lenses. Even in young children (as early as 2—4 years) the
visual acuity is generally high ($4/5$—$5/5$) except in the few cases
where a pronounced error of refraction (myopia or astigmatism)
is present. I was unable to find in my cases any confirmation
of Chavasse's rule according to which the visual acuity of the
child gradually increases from $6/12$ to $6/6$ between the ages of
2 and 5. I repeatedly found a visual acuity of $5/5$ at the age of
3 years.

In cases of amblyopia the vision generally ranged from $f/5$
to $1/10$. In a few cases it was less. As we shall see later in the
discussion of amblyopia, my cases also failed to provide any
support for Chavasse's opinion that the maximum attainable
visual acuity of an amblyopic eye never exceeds the visual acuity
which it possessed at the time at which it became amblyopic.

With timely and correct treatment the vision of the amblyopic
eye need differ hardly or not at all from that of the good eye.
Not the time at which the amblyopia began but the age at
which treatment is started is the decisive factor for the attainable visual acuity.

The significance of sight and acuity of vision — also psychologically — even for the very young child becomes apparent when we for the first time cover the good eye in order to determine the visual acuity of the amblyopic one. We shall come to this again when dealing with optical localization.

The optical localization. It is, of course, impossible to carry out a complete investigation of optical localization with these small patients. An attempt was made simply to gain some impression of the optical localization of the amblyopic eye. With very young children this impression can obviously be only very superficial. After covering of the fixing eye the child is asked to point out a figure — mentioned by its name — on one of the lower lines of the Burghardt chart which is held in front of him. If he manages to carry out the instruction, it is repeatedly observed that he points wide of the object. The pointing is always very hesitant and the movements atactic. The attitude and behaviour show that the orientation is defective. Often, in seeking the desired object, the child turns the head far in the direction of the covered eye and finally turns the trunk in the same direction. Very striking is the general impression of helplessness and misery and the child often bursts into tears at this point.

By analogy with the smile which lights up the face of the successfully treated cataract patient when he first puts on his spectacles, the expressions of these children help to show us what sight means to the individual.

The fundus oculi. After the pupil has been widened by several days' treatment with atropine, the fundus examination does not usually present any great difficulty in young children and can be carried out with precision.

Various investigators have studied the fundus of the newborn, sometimes on a large number of cases. The results found are on the whole in good agreement (Jaeger: Ueber die Einstellung des dioptrischen Apparatus, 1861; Von Sicherer & Stumpf:.Ueber Blutungen im Auge Neugeborener. Geburtsh. u. Gynaek., 13, 408, 1909; Karelitz & Vogel. Am. J. Diseases of Childr., V 50, 872, 1935; Edgerton: Ocular Observations and Studies of the New-born, with a review of the literature Arch. Ophth. 11, 838, 1934, etc.)

In general the fundus is described as pale with little pigment and very definite marking of the choroid vessels. The stratum pigmentosum retinae likewise contains very little pigment as yet. On the whole the vessels are narrow and the veins often difficult to distinguish from the arteries. The papilla is often hard to distinguish from its surroundings; in many cases (according to Karelitz & Vogel 87 %) the colour is a more or less decided grey which becomes pink in the course of the next 2 weeks to 6 months.

An important point in the literature is the occurrence of haemorrhages in the eye of the newborn, since these are believed to be connected with a possible later occurrence of a diminution of central visual acuity. Ophthalmoscopically undetectable anatomical changes in the macula following these haemorrhages may in subsequent examinations lead to incorrect classification of such an eye as amblyopic.

Both by ophthalmological and by histological means, attempts have been made to gain an impression as to the frequency of intraocular haemorrhages in newborns, including prematures. In the latter in particular, the incidence of haemorrhages was found to be very high. Edgerton observed retinal haemorrhages in 50 % of prematures and 90 % of immature infants. Among 458 normal babies he found 131 (28 %) with a retinal haemorrhage, this being in the macula in 4 cases. Most of the haemorrhages were seen along the vessels or close to the papilla in the fibre layer. From the fact that such haemorrhages were also observed shortly after birth in the retinae of 3 full-term infants delivered by caesarean section it must be concluded that obstetrical trauma is not the sole cause. Forceps deliveries, however, gave figures of nearly 60 % to 100 % for the incidence of such haemorrhages.

Other investigators report similar figures. On ophthalmoscopic examination of 400 infants on the first day Von Sicherer found 10 % with retinal haemorrhages; on histological examination of 73 cases Lachmann found haemorrhages in the retina in 5, in the choroid in 6 and in both in 2 cases. He states that percentages of 23—34.5 % have been reported by others.

On the grounds of this frequent occurrence, these haemorrhages are believed by many to be connected with disturbances of central visual acuity which appear at a very early age and are often — quite incorrectly — labelled 'amblyopia'. These 'amblyopias' are further considered to play a part in the causation of squint.

As we have already remarked, haemorrhages in the fundus occur during or shortly after birth in at least 20 % of babies. When we compare this with the figure of less than 3 % for squint, while only a fraction again of this number will show amblyopia, we see that the theoretical chance of the latter being caused by a haemorrhage — which moreover would have to be situated specially in the macula — is very small. This sequence of events is probably confined to a few rare cases. Among my 568 patients under 6 years of age were only 5 with a permanent disturbance of the central visual acuity and in all these cases the cause was ophthalmoscopically demonstrable. They were 2 cases of microphthalmus, one of which showed extensive myelin streaks; one case of coloboma nerv. opt., one of coloboma chorioideae and one of atrophia retinae pigmentosa. Granström and Magnusson found amoung 1,000 cases of strabismus convergens only 10 with an old lesion in the macula of the squinting eye; in 6 of these the cause was probably toxoplasmosis.

When we consider the fact that the transition from the congenital dissociated state of the eyes to the associated state can only be effected when the optomotor reflexes have been able to develop normally in each eye, we cannot be surprised at the occurrence of squint in cases in which this development has been disturbed in any way.

Braendstrup pointed out that children under 5 years with a unilateral cataract — whether operatively treated or not — develop a convergent squint in the great majority of cases and that this is connected with the age at which the defective vision started. He considers, rightly, that no direct causal connection exists between this squint and amblyopia ex anopsia in children, but seeks the cause in a fusion disturbance.

If extensive pathological changes occur in the fundus at a very early age, this may well prevent or seriously impede the establishment of a normal binocular coupling, with squint as a consequence.

Where these fundus changes are associated with a severe affection of the brain, as is the case with toxoplasmosis, it is obvious that strabismus can be expected in nearly all such cases. A glance at the cases in children described by Binkhorst, Granström and others will confirm this.

In view of the comparative rarity of such cases, as shown by the literature and by the writer's own cases, they will, however, have no appreciable influence on the incidence of strabismus.

General condition. Among the factors which can interfere with the natural cure of squint in early youth the general physical condition occupies an important place. Serious illness in very early life, such as digestive disturbances and especially infectious diseases, among which whooping-cough — rightly or wrongly — has a particularly bad name, may as a result of general or partial retardation promote the development of squint. This applies particularly to infectious diseases which at the same time have an injurious effect on the central nervous system (syphilis, tuberculosis, toxoplasmosis etc.).

Disturbance of mental development (mental deficiency) acts in the same direction. The fact that many squinting children are nervous and irritable, have little power of concentration and lack self-control, is further evidence of a central or functional disturbance hereditary or otherwise. This psychological lability will certainly not have been favourable for the building-up and development of the very complicated system of optomotor reactions.

In addition to these and other exogenous and endogenous disturbing factors, as much information as possible should be obtained as to the social and home conditions and the child's upbringing.

The success of the treatment to be undertaken will depend to a great extent on whether one has taken sufficient account of the reciprocal action between these factors and the individual.

During treatment there is time enough to form a correct judgement of the influence of the parents on all these factors.

Anatomical and other anomalies. Finally, our understanding of the pathogenesis of a given case may be greatly enhanced by information as to motor disturbances and anatomical abnormalities, especially as regards the form of the skull and the extremities, and also to functional disturbances such as stuttering or left-handedness.

That such a great deal of our squinting children are late in talking, walking and cleanliness points once more in the direction of a central cause, which might be the same for al these anomalies.

CHAPTER III

THE ONSET AGE OF SQUINT

THE subjects of my investigation were 894 squinting patients from my practice during the years 1945—50. They came from towns and villages in the provinces of Friesland, Drenthe, Overyssel and Gelderland and were picked out, without any special selection, from among the patients presenting themselves in my consulting room in the normal course of events. The only restriction made was that of age; for the statistical determination of the age of onset, only patients not older than 6 years were included. For the study of the manifestations the youngest patients only were used — up to 3 years of age; the criterion here was: the younger the better.

Reliable determinations of visual acuity can be carried out quite satisfactorily at the age of 3 years and often even earlier. With a little patience one can usually get an impression of the visual acuity from the child's behaviour from the age of 1 year, although the individual difference in reaction at this age is rather large.

In order to obtain reliable data it is necessary to understand the child's possibilities, to put oneself in his place and sympathize with his interests. Individual attention and insight is an essential here and is necessary to gain the child's confidence. The examination must arouse his interest and must seem to him a delightful game, so that he is glad to come back for more. Slightly older children require judicious encouragement in their attempts to carry out the instructions given. An examination which is resisted in any way by the child is practically valueless. Strangely enough, the presence of one or both parents sometimes seems to form a conditioning stimulus to such resistance. It is, of course, unnecessary to point out that a great deal of patience must be exercised to carry the examination through to the desired result.

Upon comparison of my cases with those of other investigators in the last half-century (Priestly-Smith in 1899 up to and including Nordlöw in 1942) attention is immediately attracted to the fact that my patients not only had a much lower average age but were exclusively children, so that the affection was studied entirely or for the greater part in the stage of development.

The other investigators worked with patients only some of whom were children — most of these being also much older than my patients — while a considerable percentage were adults or adolescents. In these the original picture is often completely masked by changes and adaptations.

The earlier investigators also completely overlooked the fact that the seeing and looking of older children and adults is non-comparable, both on anatomical and on physiological grounds, with that of young children and that observations made on the former group cannot be uncritically applied to the latter.

It was my intention to analyse the data both statistically and clinically; the former method, however, being used only where clinical examination should show it to be necessary.

Statistical investigation. The need for correct information as to the age at which the first signs of the affection had appeared made itself felt immediately at the beginning of the clinical examination, in the taking of the anamnesis. A certain measure of agreement on this point has only gradually been reached in the literature. In the older German literature, up to the 2nd. half of the 19th century, most writers confined themselves to stating that the condition began 'in early childhood'. The French data from this era are somewhat more precise and place the beginning between the 4th. and 8th. year. In 1903 Worth published his extensive statistical survey of 1.195 cases of strabismus convergens, in which it appeared that 75 % of these cases had started before the 4th. birthday, with the highest frequency between 2 and 4 years.

For more than 40 years this statistical survey has remained the unassailed starting point and basis for considerations connected with the age of onset; even in present-day literature it is still frequently quoted.

About 1940 Onfray wrote that strabismus convergens usually begins in the 2nd. to 3rd. year. In the Netherlands, Rochat had in the meantime placed the beginning at the age of 1 to 2 years, while Van der Hoeve had stated that in his experience strabismus convergens was observed at the age of about 3 years.

Although it thus seemed practically established that the beginning of squint had to be sought between the ages of 2 and 4 years, voices were not lacking in defence of an earlier inception. Even Worth himself remarks in his book that in 53 % of his cases of strabismus convergens alternans the condition was noticed before the 1st. birthday, while at the International Congress in Utrecht in 1899 Priestly-Smith emphatically drew attention to the very early onset of the condition and presented statistical data on 347 cases to show that 60 % had begun before the 3rd. birthday. His fellow-countryman, Richard Middlemore, had actually pointed a lifetime before him to the very early age at which squint may begin.

Congenital squint or that beginning very soon after birth has received only rare and imprecise mention in the literature. Worth and Jaensch, for instance, merely remark that it seldom occurs.

This is the more remarkable in view of the fact that ancient Egyptian literature describes a method of treatment for 'small children who squint from the time of birth'.

After the Symposium, (1943) when I subjected my cases of squint to a closer scrutiny, paying due attention to the anamnesis, I was soon struck by the fact that the number of cases in which the squint had been noticed before the 1st. birthday was by no means negligible. I then tried in various ways to get the opportunity of examining squinting babies at the earliest possible moment. Fortunately I found parents ready to co-operate, with the result that in 1946 to 1950 (15 May) inclusive I was able to find 573 very young squinters (not older thans 6 years) in my practice. Of these I examined 50 under 1 year of age, a total of 123 under 2 yr. and 344 between 2 and 4, while the remaining 56 were seen for the first time between 4 and 6 yr. In total I examined 894 cases of squint up to May 15 1950.

Upon screening it transpired that 151 cases had to be omitted from the statistical analysis on account of insufficient information as to the age of onset, while 19 had to be dropped for other reasons. Of the remaining 724 patients, 68 were found to have strabismus divergens, a condition with which I do not propose to deal here. This left 656 cases for statistical analysis.

In order to get comparable statistical data I have worked out my analysis on the same lines as that of Worth.

This means that the 2 groups, strabismus unilateralis and

strabismus alternans, in which the clinical picture of strabismus convergens presents itself are reported separately, although it transpires from the following clinical investigation that there is no essential difference between these 2 forms, while genetic studies (Waardenburg) also confirm this view on what I believe to be good grounds.

Since age differences of 1 year were too large for the placing of these very young patients in age groups, I have made a further subdivision with increments of only 6 months.

This was made possible by the greater precision of the anamnestic data now that my little patients were brought for examination at such a very early age, as the date at which the deviation had appeared was still fresh in the parents' memories.

In this way it was also possible to get an idea of the number of cases in which a squint had become evident even at birth, or perhaps more correctly, very shortly after birth. It was thus possible to add these as a separate group. It was just these 2 subdivisions, which I have not yet found anywhere in the literature, that provided such valuable information about the course, nature and origin of squint while, moreover, they provided us with the necessary data for a frequency curve.

In view of the important rôle of heredity in the pathogenesis of squint, it goes without saying that hereditary and/or familial occurrence of squint has been given a special place in the statistical surveys which follow.

Table I

Survey of 656 cases of strabismus convergens in children, arranged according to the age of onset. Age differences 1 year.

Age of onset	Strab. unilat.	Strab. alt.	Total	Percentage	Hereditary	Percentage
0—1 yr.	270	84	354	53.94	177	50
1—2 yr.	126	34	160	24.39	54	33.75
2—3 yr.	65	9	74	11.28	25	33.78
3—4 yr.	37	5	42	6.40	13	30.95
4—5 yr.	17	0	17	2.59	5	29.40
5—6 yr.	6	1	7	1.06	2	28.57
6—7 yr.	1	1	2	0.30	1	
over 7 yr.	0	0	0		0	
Total:	522	134	656		277	

From table I we see first that 354 or 53.94 % of the
total number of squinting children already had a manifest
squint at the end of their first year of life. By the end of the
second year this number has increased by 160 (24.39%), so
that 78.33% of the total number of patients already have a
squint at the age of about 2 years. This is an essential and
constant squint and is, therefore, quite different from the occasional, transitory squint which may result from the dissociation
of the ocular movements that is easily observable in *all* infants,
sometimes even for several months after birth, a phenomenon
to which we shall return in a subsequent chapter in connection
with ocular movements.

In the group up to 1 year of age it appears that the affection
is of hereditary character in 50% of cases; this percentage must
be regarded as a minimum as it is derived only from anamnestic
statements and not from examination of families. The fact that
Worth found a percentage of nearly 52 (51.78%), reckoned
on the total number of his cases, shows that our percentage
although high is certainly not too high; it is in good agreement
with Worth's findings. Squint is here regarded as hereditary when
the condition is found or reported in one or both parents or one
or both of the grandparents. The figures given in the literature
are, on the whole, much lower. Most authors report about 10 %
of hereditary occurrence (Cohn, Lagrange & Moreau, Cords,
van Duyse and others). Caesar, however, found 35% of
hereditary occurrence and Czellitzer's statistics give a proportion
of 40% from parents of whom one squints. It is obviously
necessary to define the scope of the term 'hereditary' in all
statistics.

To illustrate the degree to which squint may be at the same
time hereditary and familial in occurrence I may mention 2
families. In one the father has strabismus convergens O.S.; 6 of
his 7 children have a convergent squint (4 girls and 2 boys).
At the time of writing, however, none of his 6 grandchildren
has yet shown a squint.

In the second family the father squinted with the left eye
from birth and has worn spectacles since his schooldays. The
squint disappeared spontaneously but amblyopia O.S. remained.
Six (girls) of his 8 children developed a convergent squint.
Strange to relate, the convergent squint of one of the girls
first disappeared and then, when she was 14 years old, gradually
developed into a steadily increasing divergent squint, for which

the patient (now a nurse) will soon have to undergo an operation. Waardenburg is obviously right in his statement that squint is to a high degree a hereditary anomaly. He can hardly mean, however, — as one might be inclined to believe from the opening sentence of his contribution to the Symposium — that it is necessary to 'place heredity in the first rank of the causes of squint'. Hereditary occurrence tells us only *that* a certain case will occur in a certain way, under certain conditions, perhaps to a certain percentage, in certain circumstances or within certain limits. It goes without saying that we are left completely in the dark as to the cause which gave rise to this defined case.

Only 74 cases (11.28%) were stated to have begun between the ages of 2 and 3, while in the age group 3-4 the percentage goes down nearly to half; for 5-6 years it has decreased further to rather more than 1% and a first appearance above the age of 6 is found only sporadically.

A fact worthy of note is that heredity still remains an important factor even in these late cases.

Strabismus unilateralis was diagnosed more thans 3 times as often as strabismus alternans. It appears that the condition develops less frequently into strabismus alternans the higher the age of first appearance. My experience does not, however, confirm Worth's opinion that strabismus alternans is particularly frequent in the cases with a very early beginning. Both shortly after birth and later, the cases of strabismus unilateralis are by far in the majority. In view of the fact that transformation of strab. alternans into strab. unilateralis has several times been clinically observed with certainty, while border-line cases difficult to classify are not uncommon, the 2 forms should not be regarded as separate diseases but merely as expressions, varying in degree according to circumstances, of one and the same disease.

This does not detract from the fact that each of the 2 forms has its own character, in which the connection with its original nature is retained; this is in itself sufficient to justify its retention as a clinical entity. According to this line of reasoning one might be inclined to regard strabismus alternans as a continuation of the primitive form of seeing, as a mode of binocular vision which has not developed further than the stage of physiological two-memberedness. If this view is adopted, however, it is not permissible to use the name of strabismus alternans for an

initially unilateral squint after abolition of the amblyopia —
as is still done from time to time in the literature.

Clinically also, this condition lacks the characteristic feature
of strab. alternans, in that — as is understandable — the constant changing over is absent.

Table II

Survey of 656 cases of strabismus convergens in children, arranged
according to the age of onset. Age differences 6 months.

Age of onset	Strab. unilat.	Strab. alt.	Total	Percentage	Hereditary	Familial
0—6 months	174	53	227	34.60	123	96
6—12 months	96	31	127	19.36	54	47
12—18 months	81	19	100	15.24	35	28
18 months—2 yr.	45	15	60	9.15	19	9
2—2½ yr.	38	6	44	6.71	12	8
2½—3 yr.	27	3	30	4.57	13	11
3—3½ yr.	26	3	29	4.42	10	8
3½—4 yr.	11	2	13	1.98	3	3
4—4½ yr.	11	0	11	1.68	3	2
4½—5 yr.	6	0	6	0.91	2	1
5—5½ yr.	3	1	4		2	2
5½—6 yr.	3	0	3		0	1
6—7 yr.	1	1	2		1	1
over 7 yr.	0	0	0		0	0
Total:	522	134	656		277	217

In the above table of cases with an age-difference of 6 months
one is immediately struck by the large number of children who
had developed the affection in their first ½ year of life; this
number actually exceeding by 100 the number who started to
squint in their second ½ year.

In the subsequent 6 months the frequency of appearance
drops a few per cent, after which the figure between the ages
of 18 months and 2 yr. reaches a value which is less than half
that for the frequency in the second 6 months.

In the next 1½ yr. the frequently drops further, to ⅓ of
the above-mentioned figure, after which the frequency shows

a sudden steep descent between $3\frac{1}{2}$ and 4 yr., while after 6 yr. only an occasional sporadic case is found.

This very striking falling-off of the frequency about the 4th. year, i.e. in the period during which the organ of vision may be considered to have almost completed its anatomical development, while the correlated physiological maturation has also reached an advanced stage, seems to point to a direction in which the cause of the anomaly may profitably be sought. This idea would appear to be supported by the fact of clinical experience, that squint in an individual in the first years of growth and completion of structure, has proved in a very large percentage of cases to be an affection which tends in all respects towards regression and is curable in character.

The percentage of hereditary forms in the first 6 months is rather more thans 54%, while the occurrence is familial in rather more thans 42% in this group. By familial occurrence we understand here the occurrence in more than one child of the same parents.

Examination of non-squinting brothers and sisters in families having hereditary or familial cases showed that many of these had a labile binocular adjustment; this again emphasizes the markedly hereditary character of the affection. Statistics show, however, that after the age of 4 years cases of constant squint rarely appear in this group.

One of Worth's main arguments in favour of his theory thus loses much of its weight, since he does not state the ages.

Finally, we see from the foregoing table that in all age groups the cases of strabismus unilateralis far outnumber those of strabismus alternans and that the former type is nearly 4 times as common as the latter.

The further course and the fact that these patients are still under observation make possible to state with certainty that these patients — like all those dealt with in the statistics presented here — are suffering or have suffered from manifest strabismus convergens. There is no possibility whatever of confusion with the dissociated ocular movements which are normally seen in all children in their first few weeks or months and in the course of which, as an apparently associated movement, the true strabismus-convergens position is sometimes observed. I say expressly 'apparently associated', for a rather longer observation would have found these same eyes assuming a position of extreme divergence.

Table III

Occurrence of strabismus 'congenitally' or shortly after birth among 656 cases of strabismus convergens in young children.

Age of onset	Strab. unilat.	Strab. alt.	Total	%age	Hereditary	Familial	%age
shortly after birth	83	38	121	18.44	64	53	43.80
up to ½ yr.	91	15	106	16.16	59	43	40.56
½—1 yr.	96	31	127	19.36	54	47	37.01
1—1½ yr.	81	19	100	15.24	35	28	28
1½—2 yr.	45	15	60	9.15	19	9	15
2—2½ yr.	38	6	44	6.71	12	8	18.16
2½—3 yr.	27	3	30	4.57	13	11	
3—3½ yr.	26	3	29	4.42	10	8	
3½—4 yr.	11	2	13	1.98	3	3	

etc.' (see table II)

Table III, thus shows up especially those cases in which it was reported that the squint had been noticed from the time of birth or within the first few days. The mother is generally very positive on this point.

There is no doubt that the frequent familial occurrence leads to careful observation of the newborn in this respect.

A particularly interesting fact emerging from the above analysis is that the squint was noticed at or shortly after birth in 18.44 % of the total number of cases.

This fact is the more deserving of notice in that it has received no attention whatever in the literature of the last 50 years. The authors content themselves simply with remarking that such an occurrence is rare.

The anamnesis provided evidence of hereditary occurrence in 52.89 % of these „congenital" cases, while in 43.80 % the affection was familial.

Strabismus unilateralis occurred with more than twice the frequency of strabismus alternans.

The frequency curve shown below has been plotted from the data of the 3 foregoing tables; it is interesting in several respects and gives rise to the following considerations.

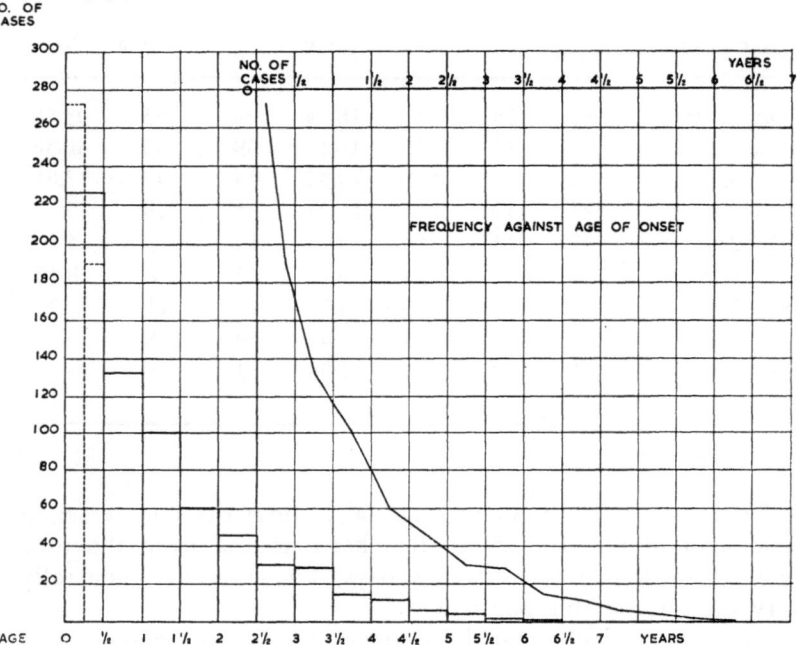

Graphic representation of 656 cases of strabismus convergens.
(Frequency against age of onset)

The dotted lines show the rectangles as they would be if the 121 cases appearing shortly after birth were included.

It is assumed that these cases developed during the 1st. 80 days of the 1st. 6 months (180 days) of life.

Since the area of the rectangle for the 1st. 6 months must be equal to the total area of the other two, the top of the rectangle for the 1st. 80 days must be at 272 (180/80 × 121 = 272) and that of the rectangle for the following 100 days at 190 (180/100 × 121 = 190).

In the first place we read from it the frequency of occurrence of squint for a given age.

In the second place the figure shows the high number of cases of squint in the first 6 months, the rapid decrease in the number of new cases in the second 6 months, the continued, but somewhat less steep decrease to the 2nd. year, then a less rapid falling off, with a few remissions, in the 2nd. to 4th. year and finally the gradually decreasing incidence until the curve becomes practically horizontal. After the 7th. year hardly any new cases of squint occur.

In the third place, the *shape* of the curve offers us a number of pointers.

Its high startingpoint might suggest *that all children are predisposed to a position of the eyes which deviates from the normal.*

The steep descent to the second half-year might then be the expression of the influence of natural normalizing and corrective factors intervening in the original course of events. In view of the period at which this descent occurs one is inclined to see in this course the growing influence of the light stimulus, resulting in the development of the function of the macula and the co-operation of the homonymous halves of the retina, this being supported by the anatomical development and finding its external expression in a continually improving co-ordination of the 2 eyes, leading ultimately to binocular (single) vision.

If this process of anatomical and correlated physiological completion had now been able to proceed undisturbed, there would be no new cases of squint appearing after the age of about 18 months; this is shown by extrapolation of the curve on the basis of this assumption.

Support is furnished for this view by the total figures of column 3 of table I. Here we see that after the 2nd. year the total of new cases falls to less than half of what it was in the preceding year.

The percentages in column 4, showing that more than 78 % of the squinting children already had this affection at the age of 2 years, point in the same direction.

At the time when, according to the literature, the condition is supposed to begin, it has in actual fact come to its natural end.

The steep descent of the curve till the age of about 1 year and its further more gradual slope, interrupted by a few irregularities would suggest, when the line of reasoning indicated above is followed, the intervention of exogenous or endogenous disturbing factors.

In this way the total number of strabismus patients divides naturally into 2 groups: one containing those who are already *primarily* affected with strabismus at about the end of the first year and the other containing those in whom the condition only becomes manifest *secondarily* under the influence of exogenous or endogenous factors.

All the investigations of strabismus so far conducted have made use of a mixed collection of cases containing representatives of both groups.

The ignorance of the very early age at which the affection begins; the more or less traditional ideas current on the subject of squint, which continued to force all investigations into the same groove; the methods employed as a consequence of such ideas (refraction, correspondence, diplopia, inhibition, scotoma, motility and position measurements) made it necessary to choose the subjects from age groups such that the patients' complete co-operation could be regarded as a certainty. This accounts for the fact that many investigators even worked with adults (Nordlöw and others), while the important group of babies and young children was regarded as useless for the collection of subjective data and was excluded as a matter of course.

When we realise further that the result of this choice of subjects was that data were collected only on old cases with all their secondarily acquired defects and compensation reactions, in which moreover a pronounced subjective tendency was undeniably present, we need not be surprised that a conflict as to the primary or secondary nature of these findings and symptoms proved to be unavoidable and continued to be waged in the literature for many years. In this way it was possible, in this conflict which was often fought on grounds of personal prejudice for lack of a physiological basis, for changes of unmistakably secondary nature to be brought forward as important causal factors and to be accepted as such by many workers.

To such a degree was it possible for the true character of the affection to camouflage itself in these complicated cases that, even to this day, the disease is identified with the abnormal position of the eyes and defined as such; in this way something that is only a *symptom,* although admittedly a prominent one, is unconsciously mistaken for *the disease itself.*

It is obvious that the nature, origin and character of the

disease can only be correctly judged and studied in the group of primary cases. The errors previously committed can be explained in this way.

In the period of development immediately following birth, during which the light stimulus is still unable or insufficiently able to make its dominating influence felt on the position of the eyes, this position is not a fixed one and is to a great extent dependent on and subject to variation with impulses received from other parts of the body (labyrinths, proprioceptors, enteroceptors etc.). This is the period of dissociated ocular movements.

The steep descent in the curve from the first till the second half year might show that this period ends at about the 6th month, although its duration is subject to individual variation.

In addition to the predisposition of all children to squint, the steep descent of the curve also shows something else, i.e. that all infants must bring with them into the world those factors upon which squint is based.

When viewed against the background of growth and maturation which predominate at this stage of development of the eye, the course taken by the curve might suggest that the predisposition with which the child is born is corrected and abolished by the growth and maturation. The causal factor with which every baby is born would thus appear to lie in the imperfections of the still incompletely developed organ.

Finally, the downward slope of the curve illustrates the natural cure of squint. If undisturbed and left to themselves these natural factors would completely have overcome the universally present predisposition in about the first 18 months.

This downward slope represents the *orthophorization.* It is the expression of the well-known striving of Nature towards harmony and equilibrium in all reactions and is, as such, to be placed on the same line as the process of emmetropization. In the one case the fluctations determine the anomalies of refraction and in the other the heterophorias.

From the age of about 1 year onwards, disturbing factors may now affect this spontaneous process of recovery, causing the definitive cure to be delayed until the age of 8 or 9 years or even later.

I am unable to confirm Worth's opinion that puberty has an influence on this process.

Every clinician is familiar with this delayed recovery from

squint, in which the child as it were 'grows out of' his squint. Such recoveries are seen already about the 4th. year and their number increases up to about 10 yr. It is worthy of remark that amblyopia, where present, seems to have only little effect on this recovery .

In contradistinction to the opinion chiefly represented in the literature, but in complete agreement with the foregoing hypothesis as to the very great tendency of squint to cure itself — on physiological grounds — I found the number of delayed recoveries with respect to the position of the eyes to be fairly large. This provides confirmation of an opinion expressed at an earlier date by De Wecker.

Once I concentrated my attention on this point I was able to collect 40 cases in a few months. These included a number of cases sent to me by medical boards examining candidates for military service, the subjects in question having severely impaired visual acuity in one eye.

A number of such cases must undoubtedly be hidden away in the group of 'cases without further detectable abnormality' which we have been in the habit of placing under the heading 'amblyopia'. Cases of this type can be recognized by the fact that the binocular bond is often still obviously labile, while in many cases the anamnesis or childhood photos will provide important evidence in support of the diagnosis.

Chapter IV

THE OPTOMOTOR REFLEXES IN STRABISMUS

A S we have seen in the first chapter, the idea that the eye movements and vision and everything connected therewith are based upon events of a reflex nature has already existed for a long time and has become a conviction for many scientists, thanks in no small measure to Zeeman's outline of its formation and nature. But even without the work of Pavlov and Zeeman, the unprejudiced investigator would find that the mere observation of what takes place in this field in the new-born and baby was sufficient to convince him that all this was due to nothing else than the play of reflex forces. It is certain that any notions of conscious mental participation in any form whatever — whereby it should be remembered that although conscious processes are cerebral processes the converse is certainly not true — or voluntary reactions, at this primitive stage of life, have to be rejected on anatomical and physiological grounds.

It is undoubtedly tempting to regard not only this primitive stage but also the further life, including all the vital manifestations of the individual, as being governed by a complex of reflex reactions occurring in accordance with laws and having achieved its coherence, co-ordination and power through a continuous succession, conjunction and reciprocal adjustment of such reactions.

This line of thought leaves no place for spontaneous, voluntary reactions; every expression is a reflex event evoked by a precedent stimulus. In this way the boundary between a voluntary and a reflex reaction disappears, in the same way that the division between unconditioned and conditioned reflexes has become more and more evidently artificial.

It is a consequence of Zeeman's theory that every reflex has at one time been a conditioned reflex and as such has had to go through a process of development. This brings with it —

and this is important in connection with squint — the possibility of disturbances in this process of development.

Even before birth the eye is subjected to influences which alter its position in the orbit. These are due chiefly to stimuli originating from the proprio- and enteroceptive fields and from both labyrinths. The eye movements evoked can be divided into monocular and binocular movements. The stimuli giving rise to monocular movements reach the eye via the monocular reflex paths; the stimuli for binocular movements proceed along the labyrinthine reflex paths (tonic cervical and labyrinth reflexes). While the monocular reflexes serve to determine the position of the eye in the orbit, the binocular reflexes have as their object the fixation of the position of the two eyes in space and can thus, in virtue of their origin and function, be regarded as belonging to the general postural reflexes.

With birth comes light which now, as a new stimulus in this field hitherto dominated exclusively by subcortical stimuli, will make its influence felt in the form of optomotor reflexes which, after a short transitional period, will dethrone the old reflexes and will soon take predominance in the determination of the position of the eyes in space and also in the orbits.

The great importance of the optomotor reflexes for the problem of squint is immediately obvious. The many other reactions (accommodation, optokinetic nystagmus, blinking reflex to a menacing movement) evoked by the optical stimulus fall outside the scope of our investigation and will not be dealt with here.

In this study we shall confine ourselves to the clinical investigation of the optomotor reflexes, examining in turn:

(1) The followingmovements in response to monocular and binocular central stimuli, that is to say stimuli which are incident on the macula or the central part of the retina in its immediate neighbourhood.

(2) The fixatingmovements upon monocular peripheral stimulation. Here we shall have to deal separately with reactions evoked from the temporal and from the nasal portion of the retina.

(3) The pupil reactions.

When considering the reactions produced as under 1 and 2 we must always be careful to note whether the movement produced is exclusively monocular or conjugated.

If we assume that the optical stimuli make use of the existing reflex paths, as mentioned in the foregoing, for the creation of their reflex arcs, we are able to distinguish three substrata, as defined by Zeeman:

(1) a binocular reflex grafted onto the vestibular reflex path;

(2) a monocular reflex grafted onto the proprioceptive reflex path;

(3) a convergence reflex, which we shall leave out of consideration for the moment (this must not be confused with the bilateral monocular adduction which usually appears at the end of the 3rd. month) and which is probably grafted onto the monocular reflexes and thus in the last analysis on the proprioceptive reflexes again.

In his theory of the development of the optomotor reflexes Zeeman also showed how various impulses may promote or hinder the development of binocular vision, according to their origin, nature and effect, and may thus contribute to the production of squint.

This line of reasoning had the obvious consequence that in the search for the cause of squint, the chief emphasis in the examination of the young patients had to be placed on the study of their optomotor reactions. This necessitated a preliminary study of the development of these reactions in normal children. The results will be reported in Chapter VI and VII.

The study of optomotor reactions in our small patients with strabismus was preceded by a determination of the nature of the squint. First we ascertained whether the condition was periodical or one of permanent squint. Then the clinical type, i.e. strabismus alternans or unilateralis was determined and finally we ascertained whether the angle of squint was constant or variable. A constant angle of squint was seldom found in our very young patients. Not only did this angle change continually in daylight, but we also noticed that many children had a much smaller angle of squint in the dark than in bright light.

It is probable that the last-mentioned differences must be ascribed to the increased gaze-tonus in bright light, the result of which will be that any defects in the binocular linking will become manifest earlier or more strongly. It is practically certain that accommodation differences can be excluded, at any rate for very young children (see also Chapter II; Refraction).

In order to avoid incorrect conclusion as to the maximal mobility in different directions, in the investigations described below, it was necessary that any motor defects found in this way should be verified. This we were able to do with the aid of acoustic, vestibular or labyrinthine stimuli (head movements; caloric nystagmus test).

The optomotor reactions of several hundred very young squinting children were now studied, first in daylight and then in the dark. We started with an examination of the ocular movements in the horizontal and vertical fields of gaze, first with monocular and then with binocular stimulation. The stimulus to be used in such tests must be suited to the child's age. For very young children one can make use of the diffuse daylight reflected by a bright surface into one or both eyes, in which case the child must be placed with his back to the window; for older children a toy (as small as possible), the shape or colour of which arouses their interest, can be used.

If one now examines the following movements of each eye separately in the horizontal field of gaze, the other eye being covered, the first examination will show a number of children in whom the following movement to the temporal side of the eye under investigation is not maximal. In strabismus unilateralis, especially where the amblyopic eye has very poor visual acuity, one usually finds some degree of limitation of abduction, i.e. a reduction of the horizontal field of gaze of the amblyopic eye at its temporal side. Very occasionally, especially in very young children (about 1 year) it seems as though there is also a slight reduction of abduction in the non-amblyopic eye. Observations on such small children — who also have a pronounced tendency to turn the whole head with (or instead of) the eye towards the light stimulus — are, however, often very difficult. Another difficulty is that one is often in doubt, with children of this age, as to whether the strabismus is alternating or unilateral, while strabismus alternans is not infrequently accompanied by bilateral restriction of abduction.

As a general rule, this method of examination does not show any abnormalities in the following movements to the nasal side or in the vertical plane in the field of gaze.

A remarkable feature in this investigation is the correctness with which the child generally manages to fix the squinting, amblyopic eye on the object.

In the search for the cause of the apparent limitation of movement we have to consider 3 possibilities: either the stimulus, or the receptor-perceptor apparatus, or the effector mechanism is at fault. It is a remarkable fact that in the history of the problem of strabismus the first 2 possibilities have always been overlooked, attention being concentrated almost exclusively on the third. The qualities of the stimulus and of the receptor are primarily and to a high degree decisive for the occurrence and effect of every reflex event.

No investigation of the optomotor reactions can be complete and no disturbance of these reactions can be assumed to exist or be further localized unless it is supplemented by tests with non-optical stimuli, to ascertain the degree of intactness of the receptor-effector mechanism.

As we have already pointed out, there are 3 possible ways of conducting this supplementary investigation.

If maximal abduction cannot be achieved at first, it is generally possible, with somewhat older children, to evoke a maximal movement to the temporal side of the eye under examination by means of an acoustic stimulus (snapping the fingers) accompanied by a certain amount of encouragement. With younger children the desired result can generally be attained by simply turning the head to and fro about its vertical axis.

If this also fails to give the required certainty, we can supplement this compensatory eye movement induced by a vestibular stimulus by the use of another test, that of caloric stimulation of the labyrinth, in which it is a simple matter to ascertain the functional possibilities of the abductor and the intactness of the reflex path, by observation of the slow phase of the nystagmus produced.

If notwithstanding the positive results of this supplementary examination it appears that a temporary or permanent disturbance in the ocular movements in response to optical stimulation is present, it is justifiable to conclude that a disturbance of the optomotor reactions of that eye exists.

If, on the other hand, a maximal abduction movement cannot be obtained by any of the means described above, it must be concluded that there is a disturbance in the function of the effector.

On the grounds of Lancaster's investigations it will still be possible, in all but a few exceptional cases, to exclude the idea of 'muscular weakness' or of excessively strong action of one muscle. As causes of a partial disturbance of function there then remain anatomical abnormalities of the muscle and ligament system or of the orbit. In none of my cases, however, have I been able to establish the presence of these with a sufficient degree of certainty.

If, however, the investigation is conducted with older or even adult patients, it will then indeed be possible to find a number of cases with such a limitation of mobility of the eyball (Scobee), although this will generally be a secondary development.

All this, however, has nothing to do with the causation of squint.

Our study of the following movements in the horizontal field of gaze of each eye separately has thus brought to light the fact that in some cases an apparent limitation of abduction exists, which must be due to a disturbance of the monocularly evoked optomotor reflexes.

In order to ascertain the behaviour of the squinting eye during following movements of the fixing eye, and if necessary also vice versa, the same tests were now repeated without the covering of one eye, i.e. binocularly. Here it will be necessary to describe the observations in each of the 2 clinical forms separately, studying in strabismus unilateralis both the behaviour of the squinting amblyopic eye with following movements of the fixing eye and the behaviour of the good (now temporarily squinting) eye with fixational following of the light stimulus by the amblyopic eye.

In strabismus alternans also it is important to examine the behaviour of each eye with fixational following of the other.

If in a case of strabismus unilateralis we now, with fixation by the good eye, move the object from the median plane thereof in a horizontal direction towards the side of the squinting eye, it will immediately be noticed that the squinting, amblyopic eye follows the fixing eye only in a hesitating manner and often stops moving long before it has reached its maximal abduction position. It thus remains at rest during the time that the fixing eye moves over the last part of its path.

If we now move the object back over the same route, pro-longing this to the farthest temporal part of the field of gaze of the fixing eye, the squinting eye remains at rest during the first part of the return journey and then gradually begins to accompany the fixing eye, continuing to follow the latter until it has reached its maximal adduction position.

If we now move the object back over the path just travelled, it will be found in very many cases that the squinting, amblyopic eye shows little or no tendency to leave the nasal angle at which it is now situated, so that at the end of the movement both eyes will be in their respective nasal angles.

It appears, thus, that in these cases the predominance of the adduction in the squinting eye is so great that the stimulus, which now acts on both eyes, fails to evoke a conjugate movement.

With respect to the angle of squint, it emerged from this investigation that this gradually increased with movement of the object towards the temporal side of the squinting eye; returned to its mean value with a return movement over the same route and retained this value more or less with movement towards its nasal side.

If one succeeds in eliciting fixational following by the more or less amblyopic eye, it will be observed that the good (non-fixing and now squinting) eye also shows an apparent restriction of its abduction, unless it has taken over the fixation from the other eye in the farthest temporal part of its field of gaze. If this has occurred, the fact will be immediately obvious in this eye as soon as the object is moved back again by the same route.

It is now behaving again as in the first test, when it was the non-fixing (habitually squinting) eye.

If the good eye is now made to squint temporarily again, by covering it in the position just taken up, it will also behave further as the habitually squinting eye. It is, however, seldom possible to create the required conditions in ordinary daylight, although in he dark room the amblyopic eye very readily fixes and follows the point of light.

Upon carrying out the same tests on following movements in cases of strabismus alternans we find some difference between the cases with a definite preference for one eye and those in which there is no such preference.

In cases with a definite preference, the other eye behaves like the amblyopic, squinting eye in strabismus unilateralis when the patient is made to fix the object with the preferred eye.

In the other cases it is usually seen that when the object is moved from one side of the common field of gaze to the other, the fixation is suddenly taken over, somewhere in the middle of the field, by the other, originally squinting, eye, to the side of which the movement was directed; this change-over occurs with a short, jerky movement in the direction of the movement of the object. The angle over which this movement takes place is equal to the angle of squint for the given direction of gaze.

If this taking-over of the fixation does not occur, i.e. if the originally fixing eye continues in this rôle over the whole field of gaze, an apparently restricted abduction of both eyes will often be observed.

From this investigation of the binocular following movements, i.e. to a binocular stimulus, it clearly emerged that in the horizontal field of gaze the eyes of our strabismus patients possessed a high degree of independence in their movements relative to each other.

A functional coupling of the 2 eyes (so-called correspondence of the homonymous halves of the retinae), for which I should prefer to substitute the clearer, and in my opinion more correct term of *binocular* junction, proved to be absent, a conclusion which had already been reached at an earlier date by Verhoeff.

It could easily be shown, by vestibular or labyrinthine stimulation, that this dissociation was not the result of abducens paralysis.

No less interesting are the results obtained by stimulation of the nasal and the temporal half of the retina of each eye in the dark room. The stimulus used was the directed pencil of rays of the 'Oculus' ophthalmoscope. From a very short distance a point of the temporal and then of the nasal half of the retina of each eye is illuminated. During this time the child is looking at the investigator, and if necessary his attention is distracted by means of a word or two. For somewhat older children this distraction is important and may even be necessary for the achievement of a result.

There are several possibilities for the reaction to the peripheral stimulus:

A. Upon stimulation of *the temporal part of the retina,* i.e. from the nasal half of the field of vision *(uncrossed path)*:

(1) in the overwhelming majority of cases a *monocular* movement in a nasal direction occurs;

(2) less frequently there is a conjugate movement in the direction of the non-stimulated eye;

(3) in a few cases no reaction occurs.

B. If, however, *the nasal part of the retina* is stimulated, i.e. from the temporal half of the field of vision *(crossed path)*, we find:

(1) sometimes a conjugate movement in the direction of the stimulated eye;

(2) in many cases, however, no reaction;

(3) in a few cases a monocular adduction;

(4) never a monocular abduction.

By stimulating the temporal portions of both retinae simultaneously or in rapid succession it is possible in some cases to induce a bilateral adduction position (strabismus bilateralis) and to cause this to continue for some time, even after withdrawal of the light stimulus.

In our young squinting children it was always possible to evoke these reactions in each eye, both in strab. unilateralis and in strab. alternans. One gains the impression that in strab. unilateralis the monocular adduction manifests itself more readily and more distinctly in the amblyopic eye than in the other eye, whereas in strabismus alternans the bilateral adduction seems to be more easily evoked. Apart from this there is little difference in reaction between the 2 forms.

Monocular adduction can be followed in older squinting children for several years, sometimes until far in the 13th year. Even after the achievement of a parallel position of the eyes the reaction usually continues to be present for a long time, according to the greater or less degree of lability of the functional junction of the 2 eyes. As already pointed out in a previous chapter, this reaction is often demonstrable also in the non-squinting children of families in which squint is hereditary or familial. Physical or mental weakness further promotes its occurrence.

A remarkable fact is that in the dark room the monocular adduction, also for the habitually fixing eye, persists for some time after withdrawal of the stimulus.

The adduction position taken up by each eye is generally maximal and is not dependent on the magnitude of any angle of squint that may have originally been present.

If, as we have assumed, the optomotor reflexes make use of the existing reflex paths, the stimulus for the conjugate movements will proceed along the vestibular reflex path and that for the monocular movements along the proprioceptive reflex paths.

From the data furnished by the above investigation it may be deduced that a connection with the proprioceptive reflex path is more readily established from the temporal part of the retina, while a connection with the vestibular reflex path is more readily

70

established from the nasal part of the retina, so that along the crossed path the connection with the binocular gaze centres prevails over that with the monocular centres, whereas along the uncrossed path the connection with the monocular centres prevails over that with the binocular centres.

On account of the great importance of the above-mentioned reactions for a correct understanding of squint and also for the ascertainment of its cause, I propose to discuss them in more detail.

For this purpose a schema of the optomotor reflexes grafted onto the monocular reflex paths and of those which give rise

FIG. I. (ROELOFS' SCHEMA)

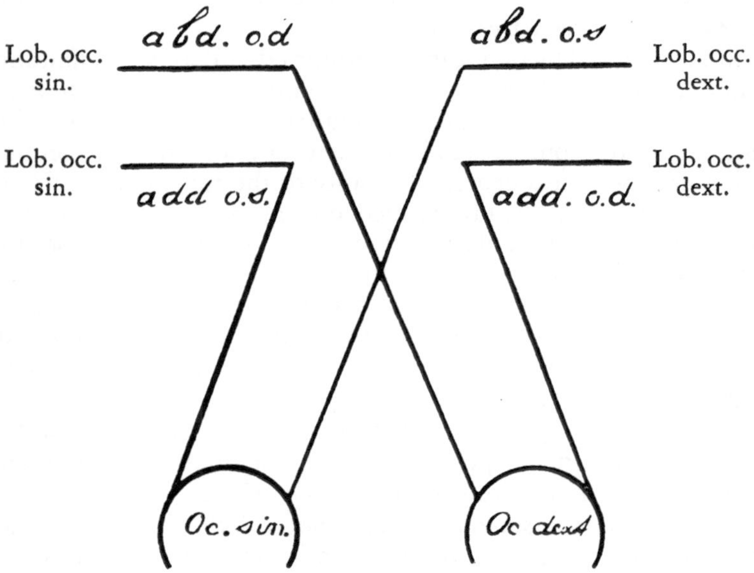

Schema of the optomotor reflexes grafted onto the monocular reflex paths. Monocular movements *(ductions)*.
Tendency to monocular vision and squint.

FIG. 2. (ROELOFS' SCHEMA)

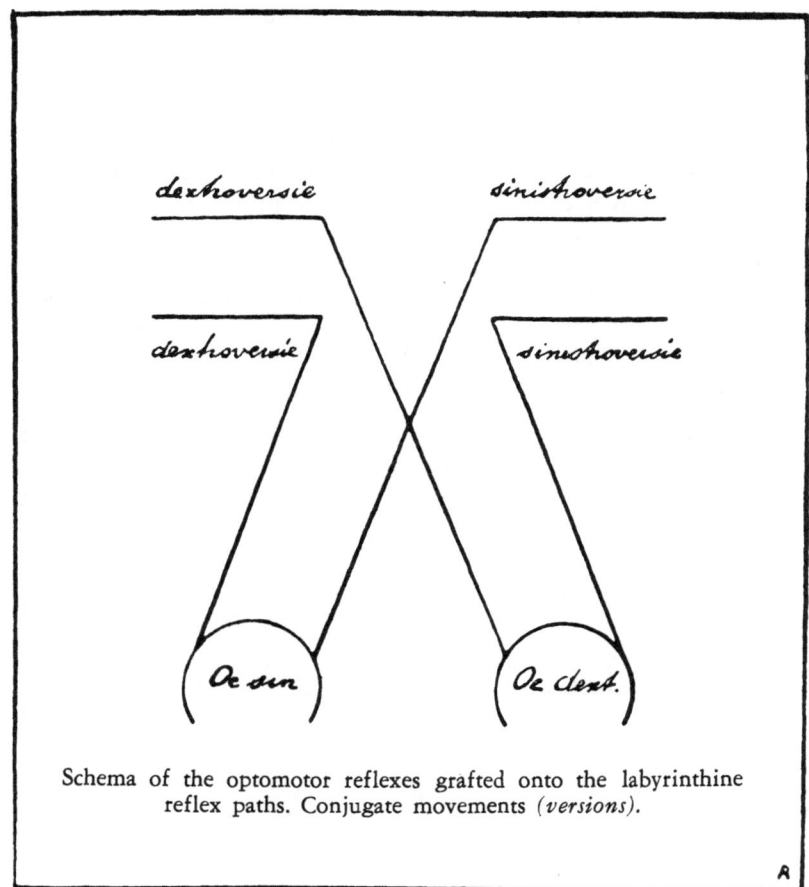

Schema of the optomotor reflexes grafted onto the labyrinthine reflex paths. Conjugate movements (*versions*).

to conjugate movements — which I have taken from some private correspondence with Dr. C. Otto Roelofs — will be found very useful. Figs. 1 and 2 represent a stage in which a complete dissociation of the 2 eyes still exists. In fig. 3 a junction has been formed, in connection with the anatomical represent- ation of the retinal elements in the optical sphere of the central nervous system, between the representatives of the retinal elements of the right and left eye (binocular junction).

re A. Stimulation of *the temporal part of the retina,* i.e. from the nasal half of the field of vision (uncrossed path).

(1)   In the overwhelming majority of cases a *monocular movement* in a nasal direction occurs.

As far as I know this is for the first time that a monocular reaction has been demonstrated.

This very pronounced dominance of the adduction reflexes is of great importance for a number of reasons. In the first place it gives objective proof that squint is based on a disturbance of the optomotor reflexes.

In the second place it explains the typical deviated position of the eyes, while in the third place it teaches us that with squint as a result of dominance of the

FIG. 3.   (ROELOFS' SCHEMA)

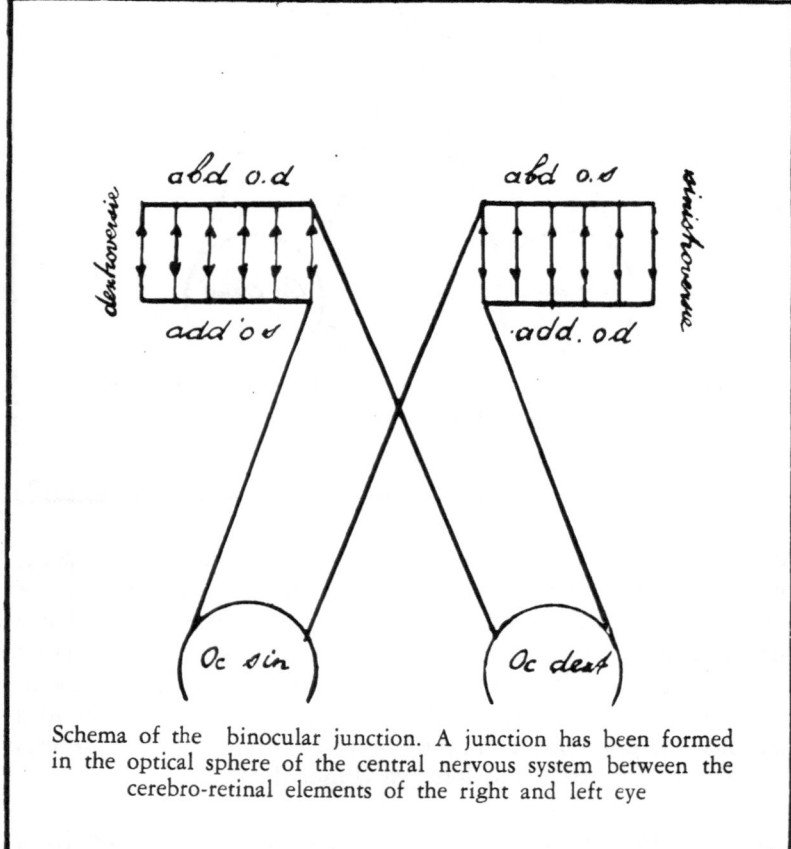

Schema of the binocular junction. A junction has been formed in the optical sphere of the central nervous system between the cerebro-retinal elements of the right and left eye adduction reflexes the functional junction of the 2 eyes has been impossible of achievement, so that the congenital dissociation of the eyes has been maintained.

Owing to this absence of binocular junction, the various binocular reflexes superposed hereon (fusion, convergence) are unable to develop and single binocular vision cannot be achieved. This stage corresponds to the condition represented in schema 1.

(2) Less frequently there is a conjugate movement in the direction of the non-stimulated eye.

Here the idea of a movement due to conduction of stimuli along the labyrinthine reflex path might suggest itself. A more probable explanation, however, is to be sought in a rather more advanced stage of binocular junction, as a result of which some of the stimuli reaching one eye can pass, via the retinal representatives of both eyes in the occipital pole of the central nervous system to the other, non-stimulated, eye. We might call this a simultaneous movement. See schema 2 and 3.

(3) In a few cases no reaction occurs.

This may happen in the first examination of a somewhat older child with periodic squint. Upon repeated examination and with distraction of the attention, however, it is always possible to elicit the reaction.

re B. Upon illumination of *the nasal part of the retina,* i.e. from the temporal part of the field of vision (crossed path).

The findings were:

(1) Sometimes a conjugate movement in the direction of the stimulated eye. Here the stimulus has grafted itself onto the vestibular reflex paths, so that a conjugate movement is evoked — which thus, in this case, is very far from being a sign of the existence of a binocular junction.

This will only be able to develop if from any point in space the same points in the corresponding retinal halves simultaneously are stimulated. A necessary condition for this is that both eyes must be correctly directed, i.e. that the monocular reflexes of each eye separately should ensure a correct position of the eye. And this last is precisely not the case with our squinting children, as

long as there exists an important degree of predominance of the monocular adduction.

(2) In many cases, however, no reaction. Apparently the
reflex arc grafted onto the vestibular reflex path is here
still insufficiently developed.

(3) In a few cases a monocular adduction. This is difficult
to explain. It may point to a potentiality for adduction
starting from the nasal half of the retina. The possibility
of an error in the technique, e.g. a reaction to diascleral
illumination, however, must also be taken into account.

(4) Never a monocular abduction.
In view of the divergent anatomical position of rest it
may be assumed that the anatomical provision for this
reflex is poor. This contrasts with the case of the adduction reflex, which is destined subsequently to develop
into the convergence reflex and for which it is to be
expected, from a phylogenetic point of view, that proper
anatomical provision will be made.
One should beware of regarding strabismus bilateralis
as evoked by a binocular convergence impulse. The way
in which it is produced, by the evocation of a monocular
adduction in each eye separately in succession through
stimulation of the temporal part of the retina, is in itself
sufficient to show that this bilateral monocular adduction has nothing to do with a binocular convergence
innervation. In my opinion the last-mentioned cannot
develop until the binocular junction has been established.

The pupil reaction to light in our squinting patients does not
call for any special remarks. It was always easily evoked, also
in and from the severely amblyopic eye.

The reaction to convergence cannot develop until later.

The study of the following and fixating movements and the
pupil reactions in young children with strabismus leads us to
the following conclusions:

(1) In strabismus there is a disturbance in the development
of the monocular and binocular optomotor reactions.

(2) The monocular adduction reflex predominates strongly
over the monocular abduction reflex.

(3) A monocular abduction reflex actually cannot be evoked
at all.

(4) As a rule no conjugate movement can be evoked from the temporal part of the retina and only sometimes from the nasal part.

(5) A dissociation of the eye movements exists, whereby the angle of squint varies widely with different directions of gaze.

(6) Owing to the predominant adduction, the monocular reflexes cannot ensure a correct position of each eye separately, as a result of which the fusion movement is disturbed.

(7) As a result of all this the binocular junction is lacking in squint.

(8) Bilateral monocular adduction is present but the binocular convergence reaction is still absent.

(9) The pupil reactions to light are undisturbed.

(10) There is only a difference in degree between the optomotor reactions in strabismus unilateralis and in strabismus alternans.

(11) This study of the reflexes made it possible to confirm the bilateral nature of the affection in an objective manner.

(12) The typical position of the eyes is the result of a disturbance of the optomotor reflexes and therefore must be regarded as merely a symptom of the disease.

CHAPTER V

THE ELECTRORETINOGRAM (ERG)

IN the search for a reason for the motor predominance of the temporal halves of the retinal, the use of the ERG immediately suggested itself. If this predominance should be due to a retinal inhibition or an inferiority of the nasal halves of the retinae, it would undoubtedly find expression in the ERG.

Our knowledge on this point is admittedly still fragmentary but it may be regarded as sufficient to give an impression of the functioning or non-functioning of the retina, at any rate of its degree of excitability and probably also of the flow of nervous impulses in the optic nerve.

Although the registration of these action potentials of the retina by the method described by Karpe is still almost unknown in the Netherlands — only the Institute for Eye Diseases ('Stichting voor Ooglijders') in Rotterdam makes use of it at present, this method has already been in use for some years in several hospitals in other countries, as an aid to clinical examination.

The fact that ERGs — several of which are reproduced below — could be obtained from 6 children with strabismus was due to the kind co-operation of my colleague Dr. H. E. Henkes, who is attached to the above-mentioned 'Stichting' in Rotterdam. In all cases the ERGs of both eyes were registered with illumination of the temporal and of the nasal part of the retina.

A difficulty with our small patients with strabismus lay in the fact that, as stated by Karpe, it is hardly possible with the usual apparatus to obtain ERGs from children under 12 years of age. However, in view of the great importance of proper understanding of the function of the retina in amblyopia, in addition to the verification of the 2 opinions encountered in recent literature, in which the inhibition of function (accepted by both parties) is held to be localized peripherally in the retina

(Harms) or cortically (Wald and Burian), we decided after all to attempt the registration of ERGs of squinting children. The fact that all the findings reported by Karpe in strabismus were in adult or almost adult patients formed a further inducement to this course.

In 6 children, one 4, two 5, one 7 and two 8 years old, ERGs were recorded with the aid of a contact lens; for purposes of comparison the ERG of a 14-year-old boy with strabismus convergens unilateralis and amblyopia was also registered. Of the 3 youngest patients, 2 had strabismus unilateralis with amblyopia and one had strabismus alternans. The 7-year-old child had strab. divergens O.S.; one of the 8-year-olds had the same, in addition to hypephoria, while the last, an 8-year-old girl, had strabismus convergens with amblyopia again. Apart from a few slight faults, due chiefly to a minor error in technique, the quality of the ERGs was reasonable to good; that of the 4-year-old child could be described as very reasonable, with only slight disturbance. Some of them follow here:

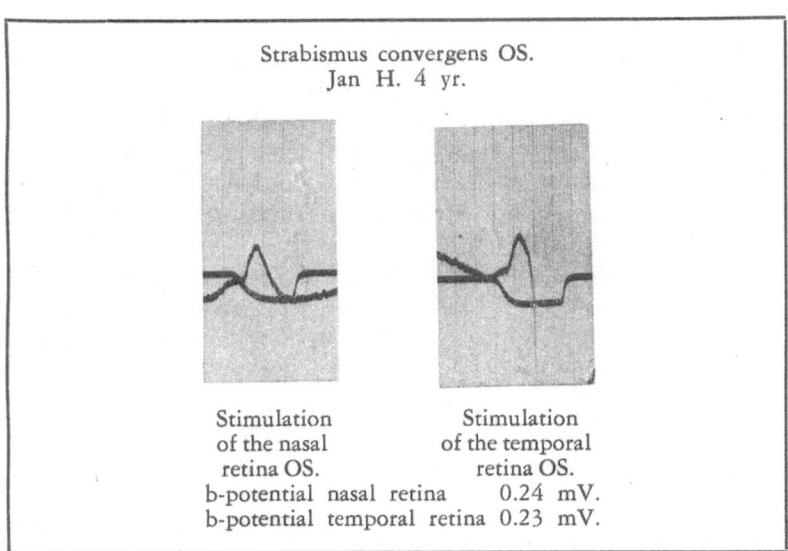

Strabismus convergens OS.
Jan H. 4 yr.

Stimulation of the nasal retina OS.

Stimulation of the temporal retina OS.

b-potential nasal retina   0.24 mV.
b-potential temporal retina 0.23 mV.

Anomalies in the ERG, i.e. departures from normal, could not be found in any of these cases. The differences between the 2 eyes were not significant and — a very important point — no difference could be found between the ERG with temporal

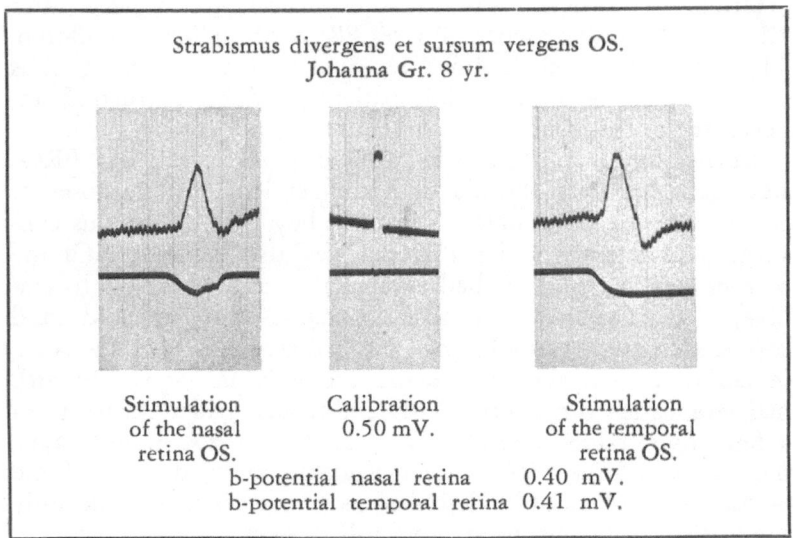

Strabismus divergens et sursum vergens OS.
Johanna Gr. 8 yr.

Stimulation of the nasal retina OS.	Calibration 0.50 mV.	Stimulation of the temporal retina OS.

b-potential nasal retina   0.40 mV.
b-potential temporal retina 0.41 mV.

and that with nasal illumination of the same eye, whether this had satisfactory vision or was amblyopic.

Although these results can give nothing more than a provisional impression, their agreement with those of Karpe, and with the findings of Wald and Burian in their investigation of threshold values in amblyopic and normally-seeing eyes, does give them a certain value.

The sensitivity to stimuli, the consequent bio-electrical reactions and the transmission of stimuli all appear to us to be normally equal and uniform in the eye with normal vision and that with amblyopia.

Since the ERG is chiefly if not entirely determined by the reaction of the peripheral retina (Karpe), this might be regarded as an objective proof of the maculocerebral character of amblyopia.

On the grounds of the normal ERG obtained both with temporal and nasal illumination it must be concluded that the perception, transformation and transmission of the stimuli by the peripheral retina is normal in both halves. The clinically-observed motor dominance of the temporal retina cannot, therefore, be due to a peripherally localized inhibition of the nasal retina but must have a central nervous cause. This provides

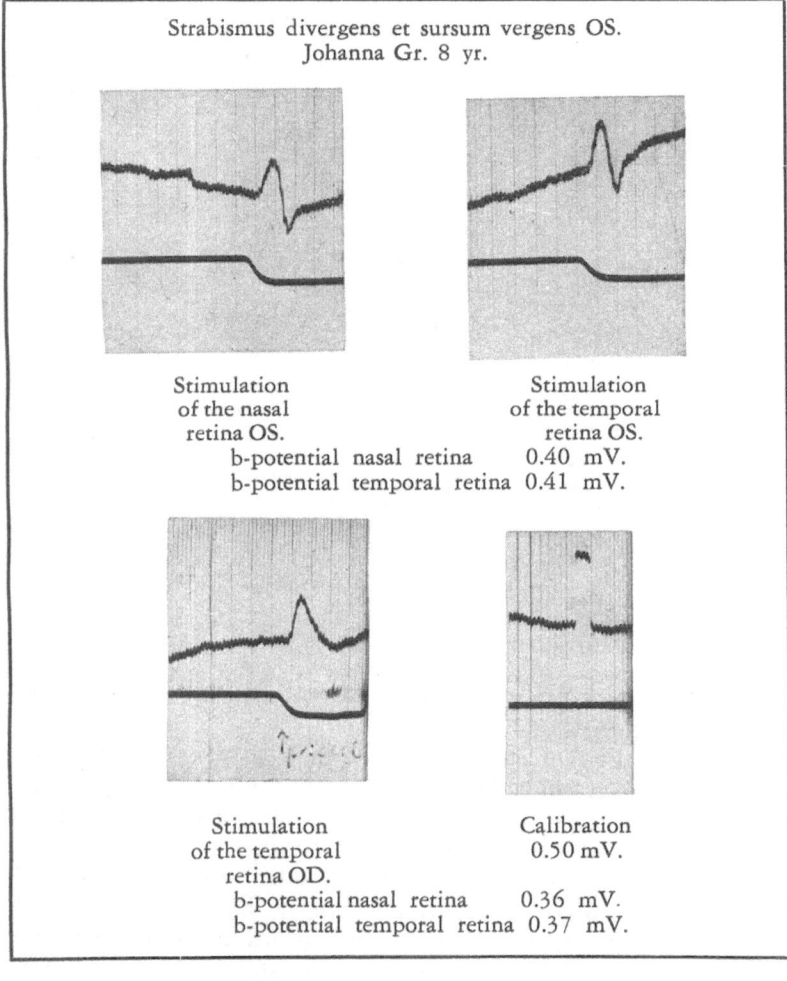

Strabismus divergens et sursum vergens OS.
Johanna Gr. 8 yr.

Stimulation
of the nasal
retina OS.

Stimulation
of the temporal
retina OS.

b-potential nasal retina   0.40 mV.
b-potential temporal retina 0.41 mV.

Stimulation
of the temporal
retina OD.

Calibration
0.50 mV.

b-potential nasal retina   0.36 mV.
b-potential temporal retina 0.37 mV.

objective support for the explanation given in the discussion of Roelofs' schema. (Chapter IV)

A similar conclusion must be drawn with respect to the inhibition of peri- and paracentral portions of the retina which has been commonly assumed to exist in the amblyopic eye, as an explanation for the absence of double images or the disappearance thereof in strabismus. All these — often highly variable — functional scotomas can be based only on a central nervous factor; an intermediate retinal inhibition, as postulated

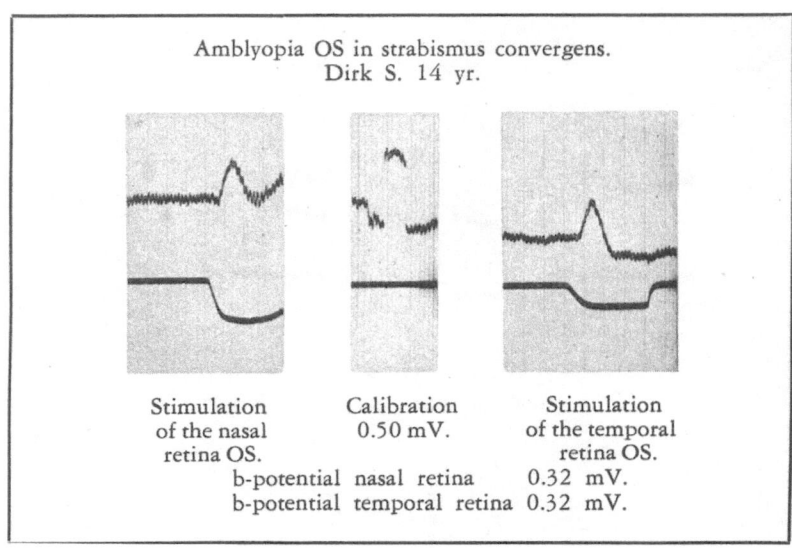

Amblyopia OS in strabismus convergens.
Dirk S. 14 yr.

Stimulation of the nasal retina OS.	Calibration 0.50 mV.	Stimulation of the temporal retina OS.

b-potential nasal retina   0.32 mV.
b-potential temporal retina 0.32 mV.

by Harms, must be regarded as extremely improbable if not entirely excluded.

It is well known that inhibition in a given cortical region may spread over the neighbouring areas. It is conceivable that a macular central inhibition might also spread to a part of the periphery in this way.

The standpoint of Harms and many others in this matter also fails to receive any confirmation from the ERG.

Of course the ERG can only give us information about the function of the peripheral organ. If it should be possible to register the electroencephalogram of the occipital pole as well, simultaneously with the ERG, we might get an impression of the transmission of the peripheral impulses and perhaps also of their velocity. Plans of this kind are now being worked out.

Further advances in our knowledge of the ERG will undoubtedly open up new perspectives.

CHAPTER VI

MYELOGENESIS RETARDATA

A careful study of the clinical picture of 'myelinic dysgenesia', also known as 'optic pseudo-atrophy of the newborn' or 'papilla grisea', to which Beauvieux drew attention
in 1926, is of interest in more ways than one.

The fact that these children are born blind, the first signs
of light perception sometimes taking several months to appear,
not only makes it possible for us to observe the birth and first
stage of development of light-perception in the human being
but also presents us, in the period preceding this, with a picture
of the state of the eye before birth and of the forces and factors
which make their influence felt at this stage.

Later also, as the function begins to develop, these factors
still remain observable for some time, until the growing influence of optical stimuli drives them into the background. As
soon as conditions permit, however, they immediately come to
the fore again, whatever the age may be. [1])

The stage during which these factors predominate is capable
of furnishing valuable information as to their origin and nature
and the level of their reflex paths, while a knowledge of the
reactions themselves will facilitate a differentiation of these from
the reactions subsequently evoked by optical stimuli.

The time and manner of appearance of the various optomotor
reactions and also the chronological order of their manifestation
can provide us, by virtue of their correlation with growth and
maturation, with important information about the condition and
the stage of development of centres, paths and connections and
thus about the possibilities available to the growing organism
with respect to seeing and looking. Every explanation that may

---

[1]) Cf. Minkowski: Ueber frühzeitige Bewegungen, Reflexe und mus-
kuläre Reaktionen beim menschlichen Foetus und ihre Beziehungen zum
fötalen Nerven- und Muskelsystem. Schweiz. Med. Wochenschr., 1922, p.
754.

be attempted on the basis of observations of phenomena associated with the organ of vision in newborns or young children must necessarily take these possibilities as its basis and starting-point.

In addition to the blindness, an important manifestation of this affection is a grey colour of the papillae. This and the gradual changes which it undergoes have been thoroughly studied and described by various authors (Karelitz and Vogel; Halbertsma; d'Offret and de Viallefont).

After his monograph published in 1926 — from which it emerges that Van der Hoeve also concerned himself with this affection — Beauvieux once more drew attention to the condition in 1947. He considers the grey colour of the papilla to be due to myelination disturbances in the optic nerve; this he has been able to confirm by pathological examination.

Beauvieux assumes that, under the influence of various factors, such as rickets, nutritional disturbances, syphilis or tuberculosis, disturbances and delays of the myelination process may occur in certain parts of the cerebrum, with the result that the alarming syndrome of blindness is produced by disturbances of conduction etc. Owing to the fact that the oculist is usually the only specialist who gets the opportunity of examining such cases, the affection is hardly known outside this branch of medicine.

Another important manifestation — which, however, is only cursorily mentioned by most investigators, with the exception of Beauvieux — is the peculiar eye movements seen in these children. In the stage of blindness both eyes wander restlessly, sometimes conjugated and sometimes entirely dissociated, in all directions, as though in search of something.

This stage of subcortically-directed eye movements will be followed by other stages and finally by that in which the movements are directed principally by cortical (optical) stimuli. Each of these stages is characterised by and is recognizable by the nature and type of the eye movements evoked in it. It is obvious that a careful study of the reactions induced and the interpretation and analysis of these — often very complex — movements into their components, according to their individual source and nature, must be of the greatest importance for our knowledge of the physiology of the ocular movements in general and of the development of the optomotor reflexes in particular. In these children Nature presents us, as it were, with a tremendous experiment, carried out under her own direction and enabling us to study in detail the nascence and development of these reactions and to compare them with those in normal new-borns. In this way we are enabled to find the track of developmental disturbances and to observe them in their course and consequences. Thus, the programme for this investigation was given to us ready-made.

During the last 5 years I have had the opportunity of examining a dozen such children and keeping them under observation for a considerable time. All showed a clinical picture corresponding to that of 'myelinic dysgenesia', but in some cases the phenomenon of the grey papilla was absent.

A case in which the grey papilla was absent had, however, already been described by Beauvieux, who points out that the presence or absence of this may perhaps be connected with the date of examination. In 2 of his first 3 cases he saw a gradual change of the grey colour to pink.

On the other hand, some of my cases showed other signs suggestive of a disturbance in the development of the central nervous system. Apart from this, the blindness, the eye movements and the course of the recovery process were identical in all cases.

From the literature one gains the impression that this affection is rather rare. Beauvieux saw 9 cases in 20 years; my figure of 12 cases in 5 years suggests that it is not so very rare.

In these 12 cases, detailed reports of which follow, the ocular movements were thoroughly studied.

I have already referred to the restless wandering movements which are evoked by idio-muscular and vestibular stimuli and by stimuli originating from the proprio- and interoceptive fields during the stage of total blindness. As we shall see in a following chapter, movements of this kind may be seen, under certain conditions, in normal newborns also.

This stage of rather slow wandering movements is followed by one in which the movements acquire a more choreo-athetosic and sometimes spastic character. Gradually the movements of the 2 eyes seem to become more combined and a nystagmoid — mainly horizontal — form of movement comes to predominate.

Then comes a period of rather fast, coarse, mainly horizontal nystagmoid, movements -- often intermittent and jerky and sometimes proceeding by jumps (cogwheel movements); these are sometimes associated but are generally dissociated and of the same restless character as the wandering movements. It appears that the movements at this stage are already elicited partially by optical stimuli.

As the influence of light stimuli becomes stronger the nystagmoid movements decrease in intensity, speed and amplitude and the eyes sometimes remain still for a moment. Certain optomotor reactions such as following and fixating movements in response to stronger light stimuli come more and more into evidence and now a fixing following also begins gradually to appear for a few moments at a time, the eyes now for the first time stopping their apparently pointless movements for an instant.

As time goes on the movements increasingly acquire the character of horizontal nystagmus; a noticeable fact is that they become continually slower and of smaller amplitude. Very gradually the eyes now come to rest and the movements are directed and determined by optomotor reactions which become continually better regulated.

In all these cases we were struck by the fact that a definite squint developed in the period in which the appearance of following (orientation) and fixating movements, and also of grasping gestures, showed that a certain power of sight was present. This squint was at first periodic but gradually became continuous, usually in the alternating form. It is worthy of note that in a few cases this squint was manifested in a bilateral form, thus clearly demonstrating the bilateral nature of the affection.

Summing up, we can distinguish the following stages of ocular movements:

(1) The stage of wandering movements.

(2) The stage of choreo-athetosic movements.

(3) The stage of pendular, coarse nystagmoid movements, sometimes jerky and with probable indications of incipient optomotor reactions.

(4) The stage of nystagmus with commencing fixating and following movements.

(5) The stage in which a resting position was reached, practically always accompanied by squint.

Stage 1 falls in the period of total blindness; stages 2 and 3 correspond to choreo-athetosic-nystagmoid movements, characterized by pronounced co-ordination disturbances. In stage 4 the fixating and following movements develop as optomotor reactions.

In the various stages distinguishable in the development of these ocular movements, a certain analogy with the phylogenetic development of the general motoricity is discernible, in which the evolution of motor functions in different animal species is linked to the appearance and coming into action of new centres and systems and in which the final complex structure — in the sense of progressive encephalization — is reached in man (Economo).

Gradually we see the originally spastic-clonic condition of the musculature in the newborn being replaced by the normal muscular tone.

From the point of view of comparative physiology, the nature of the movements suggests the following foetal stages:

(a) the stage of idio-muscular movements;

(b) the rhythmic automatic movements in fishes;

(c) the intermittent movements, alternating with moments of rest ('type striaire') or sometimes athetosic movements ('type pallidaire') in reptiles;

(d) the movements under more definitely striatal and cerebellar control in birds;

(e) the growing influence of the motor cortex in mammals.

In man we find finally the complete cortical control, although with re-grouping and further structural development of certain centres (corpus striatum etc.). [1])

It goes without saying that the above-mentioned stages of the ocular movements show a gradual, smooth transition, so that the different movements continue for some time to occur together and alternately. This makes the condition complicated and difficult to judge, at any rate in the beginning. Even with daily examination it takes many hours and sometimes weeks to arrive at any degree of understanding and acceptable interpretation of the phenomena observed in such a transition stage. In particular, it is often far from easy to decide whether we are confronted with fortuitously simultaneous and unidirectional movements or with a conjugate movement, i.e. one which is continually repeated in the same manner in response to a given stimulus, in other words, whether we can already speak of a

---

[1]) Partly quoted from M. Gourevitch: Le développement moteur et mental chez l'enfant. Report published in the 'comptes-rendus du XIe Congrès de Psychologie Paris 1937, as noted by M. Bergeron in 'Les manifestations motrices spontanées chez l'enfant'. Paris 1947.

certain degree of optomotor reaction which can be regularly evoked. It is also difficult to observe by purely clinical means whether the eye-movements in the nystagmoid state are associated or dissociated as regards chronological order and amplitude, while the accuracy of such observations is doubtful. To obtain more reliable information I made use of films, first with the normal speed of exposure, and then with an accelerated rate of exposure followed by viewing at low speed.

With increasing experience, both in these cases and in the examination and treatment of many hundreds of children with strabismus, I was repeatedly struck by the idea that these babies, with their developmental disturbances of obviously cerebral origin, in showing me the illustrative — although retarded — picture of the development of their optomotor reactions, had also given me the clue to the origin and cause of squint.

Whereas the authors mentioned in the foregoing had stressed the diagnosis and pathogenesis of the condition, the emphasis seemed to me to lie on the physiological data with respect to 'seeing', eye movements and position of the eyes which could be derived from it.

As a result of the observations on newborns not yet exposed to the action of light stimuli and on the first 'seeing' of infants, to be reported in the next chapter, this idea of an analogy between the origin and cause of the squint of the infants with cerebral developmental disturbances and the squint of 'normal' children grew in the course of months into a conviction. This conviction was further based on and confirmed by the observations on squinting children reported in the foregoing chapters, and further strengthened by the results of treatment of several hundred very young squinters. From a disease, squint now became for me a disturbed stage in a physiological process of development through which every child passes in its first year of life: the striving of Nature towards the unity-in-duality of our vision.

It became clear to me that this process of development, consisting in the orthophorization and junction of the 2 previously dissociated eyes to a functional entity regulated from one point — Hering's cyclops-eye — could only in certain cases and under certain conditions be impeded or delayed by disturbing factors.

In the course of development of every child there is a period in which these particular conditions will physiologically be present. It is the period in which the myelination of the optic pathways and centres is not yet sufficiently advanced to ensure or to give evidence of perfect conduction and transmission of optic stimuli from the peripheral organ to the occipital cortex, or from there to the motor areas. It is known that the myelination of certain parts of the optic tract is not completed until several months after birth. The stage which myelination [1]) has reached at the moment of birth decides the newborn infant's possibilities of seeing and looking. Obviously this can vary widely, from which it follows that the stage of development of the organ of vision may also vary widely, with respect to the power of sight and the optomotor reactions in the strict sense of the term, and consequently also with respect to the position of the eyes. The extreme cases are the child born 'seeing' and the child born blind, with all possible transitions in between.

In the former case the development of the optic cortex will as a rule be correlated with the myelination of the optomotor tracts, as a result of which the child will show the optical and optomotor reactions normal to its stage of development. In the latter case all reactions to optic stimuli will be absent.

It goes without saying that a disturbance may affect the cortex and the tracts to different degrees, with all the attendant consequences, while, finally, parallel with a retarded myelination of certain tracts or connections, the chronological order in which the optomotor reactions normally develop may be disturbed. All these disturbances may be banished without trace by a timely continuation of the growth processes.

Between the group of blind infants with severe disturbances, as described above, and the group of squinting children — in my opinion to be classed as having slight disturbances — there exists only a gradual difference.

It seems obvious to seek the cause of the affection in both groups in a more or less pronounced retardation of the normal development of the tracts and connections of certain parts of the central nervous system, a process to which I should like to give the name of *myelogenesis retardata.*

As long as the eye movements of these infants are not yet governed by optical stimuli and, therefore, the optical cortex does not yet participate for this purpose, the impulses for these movements can only have reached the ocular muscles via sub-

---

[1]) For a definition of this concept see p. 111.

cortical centres. According to our conception of encephalization, this process will be gradually transferred to the cortex as development proceeds.

By analogy with observations on the optomotor reactions and light perception that remain after ablation of the occipital cortex in higher mammals — also apes — we may ask ourselves whether a subcortical event must not also be held responsible, on phylogenetic grounds, for the phenomena seen at the beginning in our little patients — and hence also for those in normal infants — with respect to the very first, diffuse perception of light. (see also case III of Beauvieux, pp. 20 and 59).

In birds, vision is still bound entirely to the tectum opticum. In mammals it gradually undergoes partial transference to the cortex to an increasing degree, but in the higher mammals and the apes certain qualities of vision (light perception and spatial localization) still remain linked to the lower level. It therefore seems a logical and not too bold surmise that in the process of development in man these functions, during their passage to the higher level, may remain for some time wholly or partially bound to this lower level.

This idea fits in with the generally accepted views on the phylogenetic development, progressive encephalization and maturation of the central nervous system in the newborn.

For a long time we followed Virchow in regarding the newborn infant as a pure spinal individual. A. Collin changed this view and classified the newborn as a bulbar (medulli spinal) individual, Lesné and Richet (fils), pointing to the already observable function of the thalamus and corpus striatum as 'bulbo-médullaire-optostriée'.

Having regard in particular to the still slumbering function of the cortex, Rhenter considers a comparison with the anencephalus justifiable.

Investigators are further unanimous in considering that phylogenesis is repeated in the ontogenetic development of the higher nervous functions (Hughlings Jackson, Von Monakow, Minkowski [1]), Gourevitch), so that the succession of temporary localizations of the higher functions, such as motility, reflex reactions, sensibility, speech etc. is determined by phylogeny and ontogeny.

---

[1]) Minkowski: La Vie mentale, T. VIII de l'Encyclopédie Française blz. 10, Paris 1938. 1 re-partie: L'élaboration du système nerveux.

Continuing along this line of thought Minkowski [1]) comes, on the grounds of clinical and comparative anatomical embryological investigations, to a division of the functional nervous development in man into 5 stages. For each of these stages he describes both the clinical and the histological picture, particularly with respect to myelination which he considers to be correlated with the formation, maturation and coordination of the reflexes. All these ideas undoubtedly constitute valuable pointers. Some support for this idea may also be derived from the subjective feelings and phenomena which occur when the faculty of sight is recovering from cortical lesions (Poppelreuter, A. Fuchs, Economo and others) and also from the pupil reaction to light, which gives evidence of the power of the light stimulus to evoke motor reactions from a lower level and which never leaves us in any doubt as to its subcortical origin. If we further take into consideration the stage of development of the brain and the very variable time at which myelination of the central optic tract and occipital cortex is completed in the normal baby, (Flechsig, Pfeiffer, Le Roy Conel and others) and smooth functioning may be expected, we find, in my opinion, sufficient reason to surmise that the awakening perception of light may be of subcortical nature and that we must therefore take the 'seeing' of babies in the first weeks of their existence with a grain of salt. Even when it has become linked to the cortex, the faculty of sight will for several months have only the qualities of protopathic seeing — or at any rate of a lower cortical function, as can be deduced from various considerations, of which the stage of anatomical development reached by the macula at that time is one.

Finally, encephalization in man brings the faculty of sight entirely to the cortex, as might be shown by the absence of subcortical optokinetic nystagmus in the adult.

From the descriptions of my cases which follow, and also from the cases reported by Beauvieux, it appears that the brain can be affected in various parts or as a whole by the retarded myelination process. Psychical and psychomotor disturbances are caused thereby, in addition to associative, senso-motor and purely motor disturbances.

The general motor agitation seen in many of our patients

---

[1]) Minkowski: Revue neurologique Paris 1921 p. 1105 et 1235 q.a. Bergeron Les manifestations motrices spontanées chez l'enfant. Paris 1947.

might, regarded in this light, be the consequence of the absence
or still imperfect function of certain cortical connections (Pavlov,
Le Roy Conel III, p/146).

Motor disorders of the extremities were seen also by Beauvieux in one case, in the form of contracture of the right arm
and hand as in Little's disease, with affection of the right leg
also, but to a slighter degree. This was regarded by the neuro-
logist as a result of retarded development of the pyramidal
tracts. I also observed such disorders in 2 cases, in the form
of pes adductus congenitus, unilateral in one and bilateral in
the other case (see photos). In the first case the anomaly disappeared completely in the course of 18 months; in the second
a slight deformation of the metatarsals persisted (see X-ray
photo).

Pes adductus congenitus unilateralis in a squiting
child aged 12 months.
(Case 6. Right foot.)

It is probable that these cases may throw new light on the
possible origin of pes adductus congenitus and similar disturbances, the cause of which is still unknown (K. Bauer and W.

FIG. I

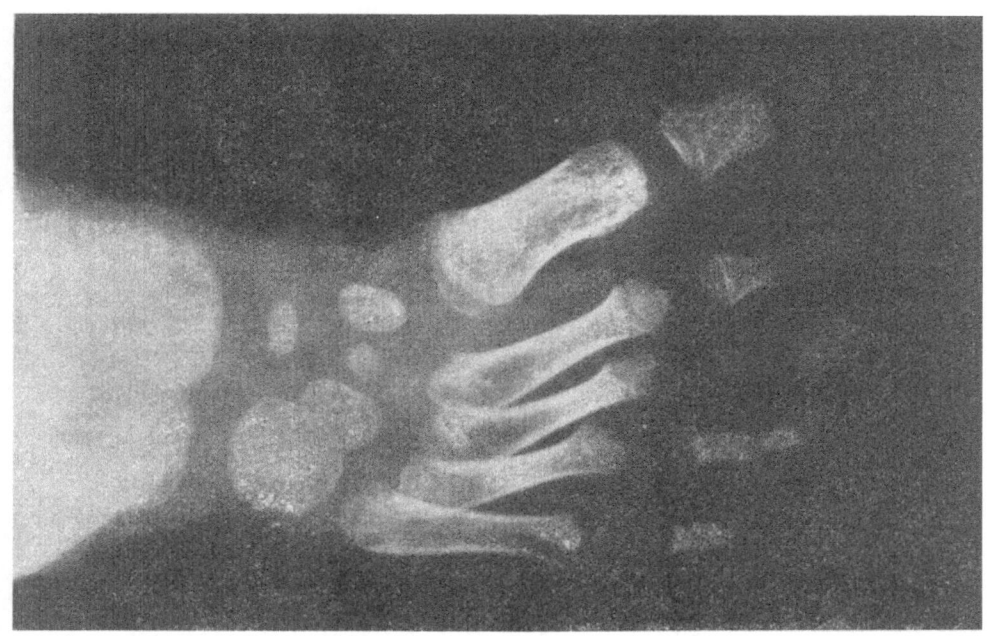

FIG. II

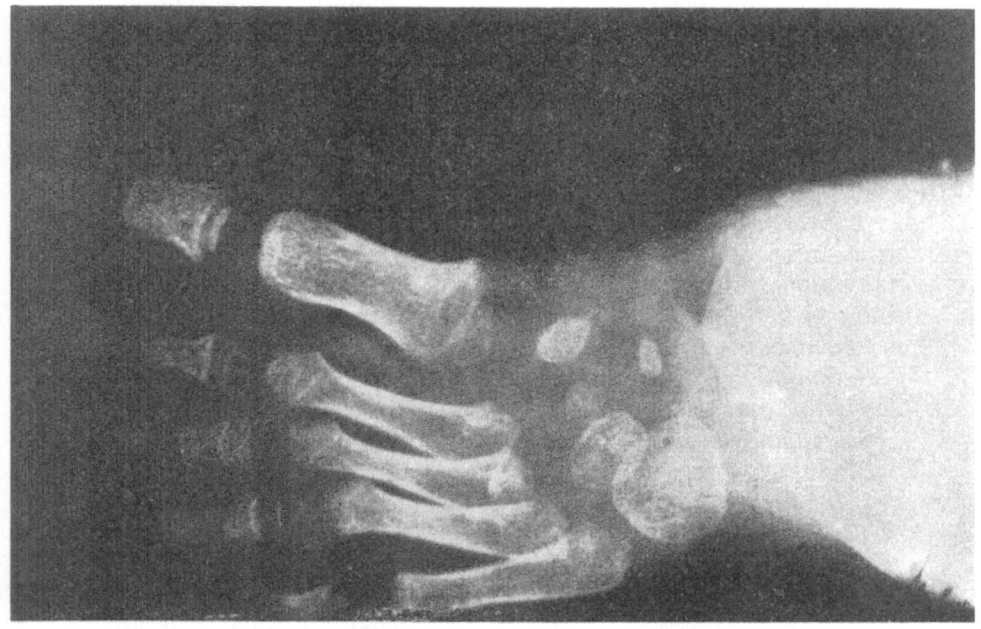

FIG. I and II: Pes adductus congenitus bilateralis in a squinting child
aged 2½ year. (Case 10)

Bode: Handb. d. Erbbiologie, 1940, III and Scherb: Fest-schrift f. Vogt 1939).

This view, in which the various affections and phenomena mentioned above are regarded as different manifestations of the retardation phenomenon, receives valuable support from the investigations of Thums, who observed 'true strabismus' in the affected partner in more than $\frac{1}{4}$ (13). of his 46 cases of Little's disease in uniovular twins. (K. Thums, Monograph. Aus. d. Gesamtgebiete der Neurologie u. Psychiatrie, VI, 1939. Zur Klinik, Vererbung, Entstehung und Rassenhygiene der angeborenen cerebralen Kinderlähmung — Littleschen Krankheit. p. 199).

All this makes it necessary to take a broader view of Beauvieux 'dysgénésie myélinique optique' and to regard it as a syndrome of which the 'gris-fer' colour of the papilla is only a symptom and one, moreover, that seems to be only temporary and not invariably present.

Despite its utility for purposes of differential diagnosis, the grey colour of the papilla is not indispensable for the characterization or explanation of the syndrome. With all the other manifestations described, it has its place in the syndrome of myelogenesis retardata. A characteristic of all manifestations of this syndrome — among which squint may possibly be included — is the possibility of complete, spontaneous recovery as a result of further growth, so that the arrears are made up.

Cases of myelogenesis retardata.

Case 1. (5.8.43). A boy aged 4 months was brought to my consulting-room by the parents, who feared that he was blind. Gestation and birth had been normal; birth at term. Although the eyes had shown continual movements since birth, the mother stated that the baby never looked at her.

Examination: Normally developed infant. The eyes were ceaselessly in movement in all directions; almost invariable with independence of each other. No fixating movement in response to light was seen out of or in the dark-room, nor was there any pupil reaction. The fundus was pale and unpigmented, with a patchy appearance here and there. Both papillae were dark-grey to black.

(26.8.43). The child was admitted to hospital for further examination. General medical examination showed nothing abnormal. Upon repeated testing the pupils, which were of rather more than medium width, were found to react very slowly and weakly to light. The eyes still showed their ceaseless wandering, dissociated movements. A trace of fixation movement in response to rather strong light was seen. The papillae were now grey.

(31.8.43). The patient was discharged with the diagnosis of papilla grisea and hope of further development of the power of sight.

(29.11.43). The patient was seen again. The eyes were much less restless and their movements slower; they now followed light readily. The pupil reaction was good on both sides; the pupils rather narrow. When direct ophthalmoscopy was performed the baby's hands grasped the instrument and he also reached out for moving objects.

The fundus showed little change.

(19.6.45). War conditions were responsible for the fact that I did not see the child again until this date. He was now 2 years old and the parents stated that his sight was very good. He had, however, been late in walking and cleanliness and had still not begun to speak. For the last year or more he had shown a squint from time to time. During my examination of him the position of the eyes was parallel. The eyes were now entirely at rest and appeared to perform conjugate movements to a monocular stimulus. The papillae were normal in colour.

(30.4.46). The child, now aged 3 yr. walked normally; his speech was still faulty. There was definite strabismus alternans with a small, variable angle of squint.

(9.10.46). The angle of squint had increased rather considerably. Under atropine refraction was R and L 2 D hypermetropia. I prescribed spectacles.

(15.7.48). The boy was brought to me again. He had stopped wearing the spectacles more than a year ago and now showed strab. conv. alternans with angle 10—15°. The total refraction was still 2 D. hypermetropia; visual acuity after correction with sph. + 2 bilaterally $^5/_5$ fig.

(21.6.49). Strab. alternans was still present but tending to strab. unilateralis O. S. Visual acuity O.D. with + 1.5 was now $^4/_5$; O.S. $^3/_5$. The child was undoubtedly somewhat backward in mental development. The fundus was normal.

(24.11.49). The position of the eyes with spectacles (sph.+2) was parallel. Without spectacles the left eye remained about 10° adducted. Visual acuity O.D. and O.S. $^5/_5$.

The papillae showed slightly pigmented edges merging gradually into the periphery. Otherwise the colour of the papillae and the fundus were normal. The boy could now talk fairly well.

Case 2: A boy aged 4 months (born 27.3.43), first seen 5.8.43. Gestation and birth normal; at term. The parents feared that the baby was blind. Its eyes had been in continual movement since birth, turning in all directions. It never looked at the light or at its mother.

Examination: A thin baby with lax musculature. The eyes showed continual wandering movements, entirely dissociated. Pupil reaction to light was absent. As far as could be judged the fundus was normal. Bilateral otitis was present. The fontanelles were normally open.

(12.10.43). Admission to hospital for observation. Both eyes still continually performed coarse nystagmoid movements in all directions. The movements were sometimes slower and sometimes faster and now and then showed a jerky character. The incidence of a light stimulus seemed to confer a rather spasmodic character on the movements; although otherwise dissociated, they were then conjugated for a moment. A strong light provoked a blinking movement; this reflex did not remain localised; sometimes the infant's whole body showed a momentary convulsive twitch. There was, thus, pronounced radiation of the stimulus. The pupils now reacted weakly to light, but slowly. The fundus showed little pigmentation. In the periphery the pigment was irregularly scattered. The colour of the papillae was normal. Discharge 20.1.43.

(21.3.44). The movements of the eyes were now much quieter and smoother in character, the pupil reaction to light was very slow.

(10.7.44). The ocular movements seemed to be less dissociated and 'seeking'. The child now fixed the source of light readily and followed this by turning the head. He did not yet grasp objects, but did utter inarticulate sounds.

(16.10.44). The eyes moved in a more associated manner. After fixation of a source of light they remained still for a time. Sometimes, however, the nystagmoid movements reappeared.

When the child's attention was attracted the eyes ceased their wandering for a moment or two.

There was now a weak pupil reaction to light.

The child was restless and fidgety. He could not yet walk and was not clean. He made speech-noises but did not grasp objects. On the whole he gave the impression of slight mental deficiency.

(8.5.45). The child could now walk and talk fairly well, but did not yet grasp things.

(12.12.45). The eyes fixed a source of light only slowly. The boy now walked and talked well and gave normal answers. The mother stated that he was not backward.

(9.7.46). The child now followed the light readily and also grasped things. I was told that he ran around without help at home, although the visually acuity appeared to be very low.

(21.7.47). Now the patient began to reach towards the light but the localization was faulty.

(17.2.48). Although the eyes were correctly directed towards the source of light, the grasping movements indicated that this was incorrectly localized. The child did not recognise objects. The mother stated that he could find his way about well at home. Only a slight nystagmus was still present. The position of the eyes was parallel.

(16.3.49). 'Degenerative ocular syndrome'. (Franceschetti).

The localization seemed to have improved somewhat but the child had no idea of distance or space. The pupil reaction to light was still slow and weak.

(5.8.49). Distinct optomotor reactions to light incident from the nasal side; sometimes a momentary reaction to a stimulus coming from the temporal side. The child now localized rather more correctly straight ahead of him. The eyes now remained quite still on fixation. Nystagmus movements still appeared in the dark. For some time the child had been squinting periodically with the right eye. Objects were still not recognized or only recognized with great difficulty. The vision seemed to be increasing slightly. The fundus was normal. The visual acuity, however, remained very poor.

Case 3. A 4 month-old girl, first seen 15.8.45. The infant was brought to me because its eyes were continually moving, it never laughed and squinted from time to time. Gestation and birth had been normal; at term. The family doctor reported that he had noticed shortly after birth that 'the eye movements were very irregular; it seemed as though the infant could not fix and was continually seeking with its eyes. Fundus examination was not possible. The pupils were wide all the time.'

Examination. An outwardly normal baby. Both eyes performed rapid, nystagmoid, apparently unconjugate movements. A coarse fixation movement in response to light was seen, but without the eyes keeping still even for an instant. The usual movement was simply interrupted for a moment to give place to a jerky movement towards the source of light. The pupil reaction to light was present on both sides, although rather slow. The rapid, coarse movements made it difficult to judge the fundus. It was unpigmented and light in colour but seemed otherwise normal as far as the papilla was concerned. The position of the eyes was continually changing and dissociated and presented an appearance of varying squint.

(20.12.45). The ocular movements were still the same, but the eyes now remained fixed on a source of light and followed this, temporarily interrupting their own movements to do so. The baby also made grasping movements.

(6.5.46). The child, now rather more than 1 year old, had shown for the last 6 months a strab. conv. alternans with tendency to predominance of the left eye, the angle of squint varied between 0 and 20°. The eyes showed fine nystagmoid movements. Walking and talking were late.

(18.6.46). Under atropine the angle of strabismus varied between 0 and 20°. The total refraction was R and L 2 D hypermetropia. Spectacles were provided: R and L sph. + 2.

(14.10.46). With spectacles the position of the eyes was practically parallel. There was fine horizontal nystagmus. Monocularly, both R and L, there was a strong optomotor reaction (adduction) to light coming from the nasal side. There was no optomotor reaction to light from temporal.

(16.5.47). The eyes were now parallel. Nystagmus as before. The child reacted with a bilateral adduction movement to a source of light held in the median plane close to the eyes. Walking and speech were still very poor and she made a slightly defective impression.

(4.12.47). The total refraction was 2.5 D.H. The nystagmus was now from time to time also vertical.

(9.11.48) The left eye showed periodical tendencies to adduction, also with spectacles. Alternating vertical and horizontal nystagmus with small amplitude. Determination of visual acuity with pictures was not yet possible.

(4.2.49). The child could now walk and talk well but was not yet clean. She was quiet and reserved. The nystagmus was now always of vertical type, but with a clockwise rotatory component. Strab. conv. alt. to 10° was periodically present.

(30.6.49). The child was now more talkative and more accessible to contact. It was, however, not yet possible to measure her visual acuity. Total refraction now H 3. Nystagmus as before.

(16.8.49). Nystagmus still unchanged. The pronounced optomotor predominance with respect to illumination of the temporal halves of the retinae was also unchanged. The eyes were parallel.

Case 4. A 6-month-old boy (born 19.10.47). First examined 30.4.48. A 7-months premature; birth weight 1,500. Stated to have been blind from birth. Had been in hospital elsewhere and discharged 31.3.48. The oculist treating him had given no hope of any sight, but the mother insisted that the baby was beginning to see a little. Both eyes had shown restless wandering movements since birth.

Examination: A healthy-looking baby. Both eyes made continual to-and-fro movements in the horizontal plane. The rhythm and amplitude were often different on the two sides. The movements showed a nystagmoid character with large amplitude and were rather fast. The pupil reaction to light was present on both sides. The patient directed both eyes towards a source of light and followed this when it was moved. The fundus was unpigmented but owing to the rapid movements the papillae could not be clearly seen.

(24.8.48). The speed and amplitude of the eye movements had decreased greatly. The child readily followed the light with both eyes and did the same with moving objects; it also made grasping movements. The papillae were dark grey; the fundus further normal but albinotic.

(16.8.49). The eyes had come to rest completely. There was slight hydrocephalus and the boy made a decidedly defective impression. He could not walk, was not clean and was very backward in speech.

He showed a periodic squint of divergent type with the left eye (to about 15°). Hypermetropia of 1 D on both sides. No monocular optomotor reaction to light with nasal or temporal incidence was detectable.

Case 5. A girl aged 2½ months (born 26.8.48). First seen 10.11.48. Had been born normally at term after an undisturbed pregnancy. The parents were afraid that the baby could not see. From birth the eyes had shown restless to-and-fro movements. For some time there had been a periodical squint, sometimes with both eyes at once. The father had left amblyopia (visual acuity $^6/_{10}$) but said he had not squinted.

Examination: Outwardly healthy baby. Both eyes performed wandering movements, mostly dissociated, chiefly in the horizontal plane. Sometimes the movements were conjugated. The eyes seemed to adopt preferentially a pronounced bilateral adduction position and remained fixed in this position for some time now and then (strab. bilateralis; see photo). Apart from epicanthus there were no further abnormalities. The pupil reaction was sluggish. On both sides there was a definite optomotor reaction (adduction) to light incident from nasal, but no reaction to illumination from the temporal side. The fundus was very slightly pigmented with distinct choroid vascular marking.

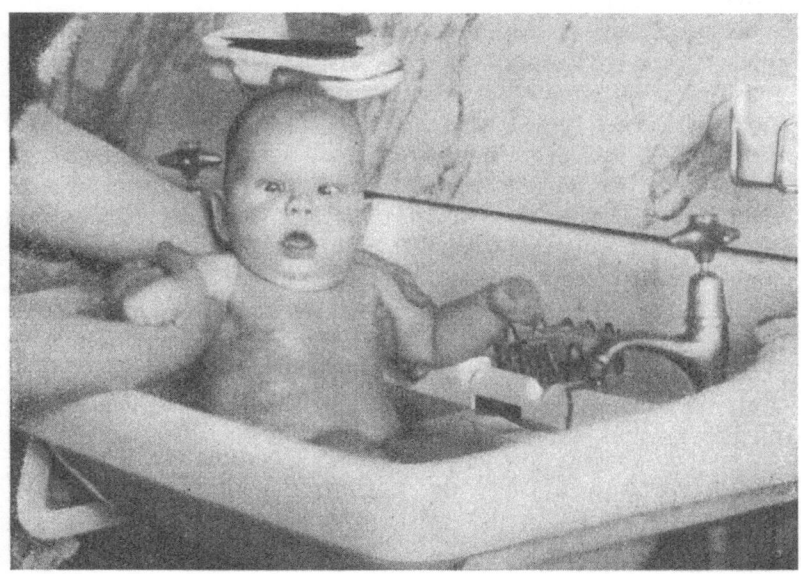

Strabismus bilateralis in a child aged 3 months with 'papilla grisea'.

The papillae were light grey with dark pigment margins.

The child fixed a source of light, the eyes then remaining still for a few moments. From time to time she showed a strab. convergens of alternating type, in which, however, there seemed to be a preference for the use of the right eye. A a rule, however, the patient reacted with pronounced bilateral adduction to light entering both eyes from the median plane.

(9.2.49). The wandering eye movements had vanished altogether. When a shining object was held before them, both eyes assumed a pronounced adduction position; this also happened when the child's attention was attracted by any other means, e.g. calling her name. There was little or no optomotor reaction to light incident from temporal, but a strong reaction to light incident from the nasal side. The pupils also reacted more promptly and remained narrow longer in response to light from the nasal side. It was not possible to get the R. or L. eye to make an abduction movement by the use of a light stimulus, whether with monocular or with binocular lateral illumination. At this stage the patient showed a frank picture of bilateral abducens paralysis. The faulty diagnosis was prevented, however, by the movements in all directions seen in the early stage and by the unmistakable abduction movements in the caloric nystagmus test. The fundus was still light in colour. The papillae were still rather greyish, with a pigmented margin. The child was not very active but did make grasping movements. She seemed to be quite normal mentally.

(13.7.49). At the age of 10 months the child had become much more active. She grasped objects of all kinds and laughed; could already stand in the play-pen and showed a very decided desire for movement.

Spontaneously, and also on looking at an object held before them, the eyes continually returned to the bilateral adduction position (strab. bilateralis), as a photo (taken in the 7th. month) shows.

A preference for the right eye became increasingly evident, however, the left eye then showing an angle of strabismus of about 30°.

If, when the eyes had assumed the bilateral adduction position, an object held in front of the patient was moved to the left of the field of vision, the left eye remained in its adducted position. If the object was then moved to the right side of the field, the right eye usually moved from its adducted position to a point just past the middle position. Upon movement of the object back the way it had come, the right eye gradually returned to its former adduction position. The left eye remained immovable all this time and kept its original position. Only occasionally, however, did I fail to entice the right eye out of its adducted position in this way. It thus appeared that the two eyes were still almost entirely dissociated. There was still a strong opto-motor reaction to light incident from the nasal side of both eyes.

(16.8.49). The child could already walk and was very quick in everything. She showed a pronounced urge to movement and motor restlessness. Strab. bilateralis still appeared occasionally. Now, however, the unilateral type predominated, with O.D. as the favoured eye. The dissociation of the eyes seemed to be less marked with movements in the horizontal field but was still very noticeable. Upon illumination from temporal (nasal half

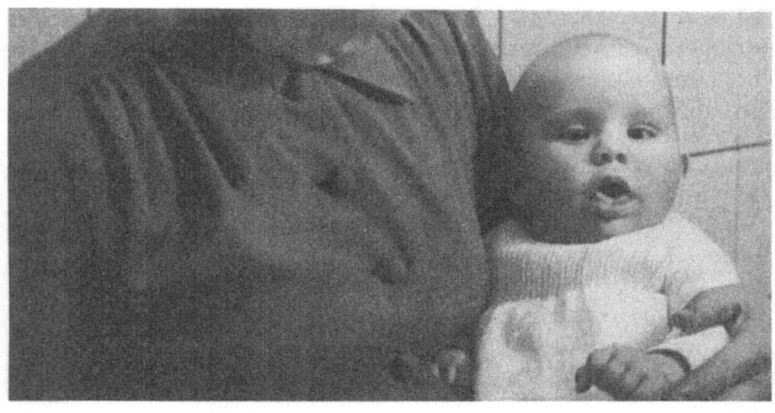

Strabismus bilateralis in a child aged 7 months.

of the retina) a conjugate movement of both eyes towards the side of the stimulated eye could be elicited, showing that the reflex arc grafted on to the labyrinthine reflex paths was now also completed. In this way the dissociation had become a partial one. This was further proof that the lateral rectus muscles had a normal function in this case. The black margins round the papillae were still unchanged.

Case 6. A girl aged 3 years; first examined 29.6.49. A 7 months premature. The pregnancy had been undisturbed and the birth normal. The child had been blind from birth. Both eyes had continually wandered round as though seeking something. There had been no reaction to light. After about 6 months the ocular movements had become quieter and it had seemed as though the baby was beginning to see; she sometimes turned the eyes towards a light. At the age of one year she had begun to grasp objects. She had first walked at 2 yr. and still did not speak. About 18 months previously a squint had appeared. A photo from this time (17 months) showed a distinct strab. conv. O.D. The angle of strabismus was still very variable; sometimes the strabismus was bilateral. Squint was hereditary in the family. At the age of 2½ yr. the patient had been provided elsewhere with spectacles, (S—10) with which she saw well.

The right foot was deformed. The mother said she had first noticed this 6 months after birth. Enlargement of the heart had recently been diagnosed. The child had never been seriously ill.

Examination: A healthy-looking but restless child, with an exaggerated urge to movement. Epicanthus.

The position of the eyes seemed normal; there was no nystagmus. There was a marked optomotor reaction (adduction) to light incident from nasal (bilateral). No reaction to light from the temporal side.

(26.7.49). Sciascopically, myopia of 12 D on both sides. Papillae normal in colour; myopic crescents alongside. Pigmentation of the fundus normal.

Measurement of the visual acuity with glasses was impossible because the child's attention could not be fixed even for a single instant on the pictures of the reading chart. Spectacles (Spher. —11) were prescribed.

The right foot showed the picture of pes adductus congenitus; the anterior half of the foot was bent medially (see photo). The left foot and the upper extremities were normal.

(16.8.49). The position of the eyes was parallel with and without spectacles. The mother stated that the child saw very well. A bilateral adduction movement was seen upon fixation of an object held close to the eyes. The optomotor reaction was unchanged.

(12.12.49). The child's movements were noticeably jerky. She was beginning to speak. There was still no optomotor reaction to a light stimulus from temporal. She directed the eyes with a prompt bilateral adduction movement to very close objects.

Pes adductus congenitus unilateralis.

Case 7. A boy aged 5 months (born 28.1.49); first seen 8.7.49. Gestation and birth normal; 14 days over term. The parents were of the opinion that the child had been completely blind for about the first 6 weeks. From the beginning the eyes had been in perpetual wandering, 'seeking' movement. Not until the age of 2 months had the baby first turned its eyes towards the light. The grandfather had nystagmus; there was no strabismus in the family.

Examination: A healthy-looking baby with totally albinotic iris. Even with reflected light the completely transparent character was immediately noticeable. In the ophthalmoscope the projections of the ciliary body were sharply visible. The eyes made rather fast, coarsely nystagmoid movements, mostly dissociated. The movements were chiefly in the horizontal plane. In the course of these movements the infant sometimes fixed a light and followed this for a moment. Both pupils reacted sharply to light. The fundus was unpigmented. On account of the rapid, often jerky movements, the colour of the papillae could not be ascertained with certainty; it seemed, however, to be normal.

(13.7.49). The total refraction was bilaterally 5 D. hypermetropia. The fundus was albinotic but otherwise normal. Bilaterally there was a prompt optomotor reaction (adduction) to light incident from nasal. Light incident from temporal did not evoke any reactive eye movement.

(10.8.49). The child had become much more lively; it smiled and grasped things and followed the source of light well. There was now horizontal nystagmus of large amplitude. The movements had become somewhat slower. For some time the child had been squinting with the left eye.

(31.10.49). The child was developing well. He had stood at 7 months and was already beginning to speak. He saw and grasped at every objects around him. The mother had noticed a definite slowing-down of the horizontal eye movements. The movement was not maximal in either direction; sometimes small and sometimes larger. From time to time the eyes stopped for a moment in the right temporal angle. The movements sometimes seemed to be conjugated; sometimes, however, they became dissociated again for a short time. It was often difficult to make a correct observation on this point.

(19.12.49). The eye movements were getting slower but were otherwise still of the same type. The iris appeared less diaphanous than formely. The papillae could now be satisfactorily observed and showed a normal colour.

At this time there was definite strab. conv. O.S. with variable angle.

Case 8. A 2-month-old infant (born 15.6.49). First seen 12.8.49. Gestation and birth normal; at term. The parents believed the child to be blind because it never looked at the light or at its mother. The eyes had been in perpetual movement since birth. From time to time the baby 'squinted horribly'. The last few weeks, however, this had not been quite so bad.

Examination: A well-developed baby with both eyes in continual, non-co-ordinated movement. They wandered, sometimes gradually and sometimes with sudden jerks, in all directions. The iris was completely transparent; the entire sclera also showed red translucence with incident light in the dark-room. The pupils, 2—3 mm. wide, reacted sluggishly to light. The fundus could not be examined on account of the ceaseless movements. Admission to hospital for observation.

(15.8.49). The patient showed general motor restlessness. The pediatrician reported as follows: 'The infant is restless and its movements are certainly more jerky than is normal at that age. It does not give the impression of being definitely defective. The musculature seems satisfactory but the skin is rather rough and shows a tendency to seborrhoea. I think the reflexes are normal. I should like to have another look at this child'.

(16.8.49). A conjugate movement could be elicited by illumination of both eyes with strong light. Otherwise the patient did not react to incident light (including illumination of the temporal retina).

Both eyes still displayed jerky, nystagmoid movements — often clearly dissociated — in all directions. Immediately after waking the eye movements were always less vigorous.

The compensatory movements and rotatory nystagmus were normally present, in addition to the cervical reflexes.

The infant continually made coarse, atactic movements with the head, arms and legs, as in athetosis. The movements were so violent that it hurt its fingers against the sides of the cot. General medical examination and X-ray examination failed to reveal any abnormalities.

(18.8.49). The movements were quieter. The eyes, too, moved less and showed slow, mainly horizontal, nystagmoid movements. The pupils were now pin-head sized in daylight; the eyes were directed for a moment towards a strong light.

(19.8.49). The jerky movements of the head and limbs were less frequent and much less violent. The face — which at first had been rather mask-like — showed more expression. There was definite fixation of a strong light-source and the eyes then remained still for a moment. The nystagmus was slower and less dissociated. With the aid of a lid-holder and fixation forceps I now managed to examine the fundus. It was markedly albinotic; more peripherally it was difficult to distinguish between the retinal and the choroid vessels. The left papilla was normal in colour and the right one very light grey; the difference was slight. Even with reflected light the iris was translucent and the projections of the ciliary body showed through it. Grasping movements had not yet been seen.

(21.8.49). Patient discharged.

(30.8.49). The infant was quiet when reexamined. During breast-feeding the eyes stopped moving. It now fixed a weak source of light, also in daylight. The horizontal nystagmus of large amplitude had now been replaced by small, irregular, horizontal movements.

(1.11.49). The mother had also noticed that the eye movements were becoming steadily slower and smaller. Now and then the eyes kept still for a moment and followed the light or an object held in front of them. The movements were mostly un-co-ordinated. For the last 6 weeks the child had periodically squinted, sometimes bilaterally. A strong optomotor reaction of adduction in both eyes could be evoked by illumination of the temporal retina. There was no reaction to illumination of the nasal retina.

For the last 4 weeks the child had had violent fits of screaming, during which it became cyanotic. It jerked its arms and legs and could not be calmed by any means whatever. This condition sometimes lasted several hours.

(20.12.49). The pediatrician stated that the reflexes were exaggerated. 'The child is still rather restless and unable to keep still; now and then it feels rather spastic to the touch.' No pathological reflexes were found. The eyes now moved slowly and generally synchronously, now again with greater amplitude.

The child followed all objects with the eyes and grasped at them. It appeared to be mentally normal.

(7.2.50). The child was fairly quiet and no longer had screaming fits. On the whole it reacted but little to stimuli from its surroundings and had little contact even with the moher. However, it followed light and objects readily, in daylight. Sometimes the eyes remained still for a moment during fixation of a source of light; sometimes, too, the child squinted — occasionally even bilaterally. The optomotor adduction reflexes were very marked; occasionally there was also a conjugate movement in response to illumination of the nasal half of the retina. I was told that he reached out for toys at home.

(21.2.50). The infant was sent to hospital by the skin specialist, on account of a very extensive eczema. This gave me the opportunity of observing it daily for nearly 3 weeks.

There was now a fine nystagmus, in addition to coarse nystagmoid movements which were frequently unconjugated. Periodically there was a squint with the left eye, with a variable angle of strabismus — sometimes even to 20°.

(22.2.50). During sleep the eyes performed the well-known wandering movements, *but there was then no nystagmus.* It appeared, thus, that a light stimulus was necessary for this.

(25.2.50). An attempt was made to film the squint and the ocular movements. In addition to nystagmoid movements there were also conjugate pendular movements.

(26.2.50). The squint was becoming more and more constant, but the angle varied continually between 0 and 30°.

(1.3.50). For the last week it had been possible to speak of a constant squint with the left eye. When the right eye was covered the child turned its head away in order to get this eye free again, whereas it cheerfully tolerated covering of the left eye. The ocular movements were becoming steadily slower and smaller. Sometimes the eyes remained still for a moment.

(3.3.50). Another film was taken. Predominant strab. convergens O.S. to 20°.

(6.3.50). Constant strab. convergens O.S. Slow nystagmoid movements in the horizontal plane.

(8.3.50). When the right eye was covered the child snatched its head away so quickly as to give rise to the suspicion that the left eye was becoming amblyopic.

(10.3.50). The squint with the left eye had visibly decreased. The eyes were sometimes parallel again. Periodically there was now also a squint with the right eye. The squint had, thus, acquired an alternating character. Covering of either the right or the left eye was tolerated without turning away the head.

(13.3.50). Discharged. The eczema had practically cleared up. The back and hip muscles were so weak that the child could not sit up.

(8.8.50). I now saw the child again for the first time. It was beginning to speak but could not yet sit up. It grabbed every object held in front of it. It squinted periodically and alternately. There was a marked predominance of adduction when the temporal half of the retina was illuminated. Illumination of the nasal half of the retina was followed by a conjugate movement. There was slight nystagmus with slow, small movements. The child appeared to be mentally normal.

Case 9. A 6-month-old girl (born 4.3.49). First seen 1.10.49.

Gestation and birth normal; at term. For the first 14 days it had been yellow. The left arm and leg had been rather weak for the first 4 to 5 weeks. A finger placed in the palm of the left hand was not grasped. The eyes had wandered from birth. The family doctor wrote 'At first the infant did not seem to look at anything. The eyes certainly moved to and fro, and in a fairly normal way, but the infant did not fix at all.'

Examination: A very fat, sturdy child. Birth weight 4.5 kg. There were no longer any wandering movements and no nystagmus. The ocular movements were dissociated, however, while a squint appeared from time to time. In the dark-room the baby sometimes followed the light and sometimes did not. Fixation of a light-source was also inconstant. There was a prompt optomotor reaction of adduction to illumination of the temporal retina; no reflex could be elicited from the nasal retina. The fundus was normally pigmented.

The papillae were rather greyish and had a pigmented margin. The child did not yet make grasping movements.

Case 10. A 4-year-old boy (born 8.9.45). First seen 12.12.49. For some time after birth he had seen very poorly or not at all. At that time both eyes were in perpetual movement. Gestation and birth normal; at term. Two weeks after birth the mother discovered that both feet were deformed; the child was then treated by a surgeon. Convulsions developed at the age of about 6 months and soon after this he began to squint. For 3 years he had been under treatment for this elsewhere, with atropine drops. This had given no result. The child was nervous and liable to fits of temper.

Examination: A rather mongoloid type, with open mouth and epicanthus. Had been late in walking, was not yet clean and spoke only a few words. Squinted alternately, but with a definite preference for the left eye. The angle of strabismus was variable, with a mean of about 25°. I was informed that the eyes were often parallel on waking.

There was definite dissociation of the eyes with movements in the horizontal field. Each eye gave a marked optomotor reaction with strong adduction in response to light incident from nasal. No reaction was observed when the light was incident on the nasal half of the retina. In the dark-room a strabismus bilateralis was seen from time to time. The fundus was rather strongly pigmented. The papillae, with their pigmented margins, were normal in colour.

Both feet showed the picture of pes adductus congenitus, with the anterior half of the foot bent medially (see X-ray photo pag. 91). The ankles seemed to be thicker than normal. The child walked well, but I was told that he often fell down.

(27.12.49). Strabismus, predominantly with O.S., was also seen under atropine. The angle now remained at about 25°. Sciascopy showed bilateral hypermetropia of 4.5 D. Spectacles with spher. + 4.5 were prescribed.

Case 11. A girl aged 4 months (born 18.9.49). First seen 17.1.50.

Gestation and birth normal; 14 days early. The baby had never looked at the light or at its mother and she feared that it was blind. Since birth the eyes had been in constant movement — rather rapid and in all directions. All possible positions, from strab. conv. to strab. divergens were passed through, sometimes bilaterally.

A week earlier the pediatrician had noticed that the infant reacted to strong light with a blinking movement.

Examination: A small, healthy, very lively baby. The upper trunk was short and the fingers long and gracile.

The eyes moved restlessly in all directions, jerky movements in an upward direction being noticeable. In both eyes the movements were wandering rather than nystagmoid and mostly dissociated. Illumination with strong light did not evoke any fixation movement. The pupils did not react to light and were 3 mm. wide. The cornea reflex was very sharp, with marked radiation over the whole body. The child was admitted to hospital for observation.

(24.1.50). Fundus examination. The right papilla was light grey and a temporal, crescent-shaped portion of it was dark grey The left papilla showed a nasal, fairly large crescent-shaped area which was grey, while the rest was pale pink. In both eyes the fundus was pale and practically without pigment. The macula region had a granular appearance.

The eye movements showed little change.

(26.1.50). There was a trace of light perception; the child blinked when a strong light was shown.

(1.2.50). A trace of fixation movements was seen. Both pupils reacted fairly readily to incident light. The eye movements were slower and smaller.

(5.2.50). A strong source of light (Sachs' lamp) was now definitely fixed with a conjugate movement. Sometimes the eyes then kept still for a moment. The baby smiled.

(7.2.50) The pupils reacted promptly to light.

(8.2.50). A strong light was now definitely followed in all directions, although the movements were not yet smooth, but were sometimes interrupted, while now and again the eyes deviated sideways for a moment (cogwheel movements). Monocular reflexes upon temporal or nasal illumination of the retinal halves could not yet be evoked. There was no winking reflex to menace.

(10.2.50). There was a distinct monocular adduction reflex bilaterally upon stimulation of the temporal half of the retina with strong light. Illumination of the nasal half had no effect.

The eye movements were now more nystagmoid and chiefly in the horizontal plane.

(14.2.50). Nothing new. Patient discharged.

(14.3.50). The eyes now moved slowly to and fro in the horizontal plane and sometimes stood still for an instant. They were now also directed to the light of the ophtalmoscope. On both sides there was distinct adduction on illumination of the temporal half of the retina; stimulation of the nasal half gave no reaction. The child squinted periodically with the left eye. The motor restlessness had decreased greatly; the child smiled and made grasping movements. It repeatedly pressed the backs of the hands against the eyes. (Symptom of Franceschetti.) There was still no blinking reflex to a menacing movement.

(4.4.50). The eyes kept still much of the time and only performed occasional co-ordinated movements to and fro (sometimes also upwards).

Strab. convergens O. S. periodic. Distinct bilateral adduction reflex. No reaction to nasal illumination. Very sluggish pupil reaction to light.

(25.4.50). Slow, pendular eye movements, mostly conjugated. Fixed the ophthalmoscope light well but followed it poorly. Grasped at the ophthalmoscope but not at objects shown in daylight.

Alternating right and left squint, sluggish pupil reaction.

(16.5.50). The child often pressed both hands to its eyes. Alternating squint. From time to time the position of the eyes was totally dissociated, with one eye turned upwards and the other temporally. The following movement was still weak. The pupil reaction was sluggish and often indefinite, even with central illumination.

(27.7.50). At intervals the eyes kept quite still. Prompt adduction reflex. Very occasionally there was a week conjugate movement observable in reaction to illumination of the nasal half of the retina.

Case 12. A boy aged 6 months. First seen 4.5.50.

Gestation and birth normal; 14 days early. From birth the eyes were in perpetual wandering movement; the baby did not react to light and never looked at its mother. Even now it hardly did so. At first the parents had believed it to be blind. From the 4th. month, however, improvement appeared and for the last 3 weeks the child had (I was told) even made grasping movements. As the abnormal eye movements persisted, however, the pediatrician considered examination by an oculist necessary.

Examination: A normal, healthy baby whose eyes moved almost ceaselessly to and fro in the horizontal plane, sometimes conjugated and sometimes definitely unconjugated. Sometimes they remained still for an instant; the size and speed of the movements were variable (had been greater). From time to time there was a suggestion of squint with the R. or L. eye. In daylight the child followed objects with eye and head movements.

In the dark-room, both eyes were directed to the source of light and followed this in all directions. On both sides there was a strong adduction reflex to illumination of the temporal half of the retina.

Only very occasionally was illumination of the nasal half responded to with a conjugate movement. The blinking reflex to a menacing movement was present.

Both pupils reacted fairly readily to light. The fundus was pale and the papillae normal in colour.

(26.5.50). The eyes moved more slowly with small amplitude, often not associated. Sometimes they remained still for a moment as though fixing and the child reached for an object held in front of it.

With sideways glancing movements the eyes could also be kept still for a moment, after which a few coarse nystagmoid jerks generally followed.

(24.7.50). Both eyes often remained still; only now and then did they oscillate simultaneously to and fro. The child followed objects with both eyes, grasped them and smiled. Now and then he squinted momentarily with the R. eye. There was strong predominance of the monocular adduction reflexes.

He seemed mentally normal. Had recently had screaming fits at night.

In all these cases we are concerned with children whose central nervous system at birth had lagged behind the rest of the body in growth and development, often to a considerable degree. This freak of Nature presented us with the opportunity of studying the maturation and unfolding of these important parts (a period of development otherwise entirely hidden from us) by observation of the phenomena belonging to or correlated with it. In this way we were able to gain some understanding of the normal anatomical and physiological development of these parts and of the chronological order in which their various functions normally manifest themselves. With the aid of our knowledge as to the anatomical conditions existing at this time, it is possible in this way to gain an impression about the pathways along which the different stimuli can reach the central parts, while the presence or absence of reflex reactions tells us whether the connection between peripheral and central organ has been established and, if so what reactions the central organ is capable of at that moment. Comparative anatomy and physiology are often indispensable, too, for a better understanding of the chronological order of appearance of the various phenomena.

By analogy with Bolk's views on retardation — which we might call total retardation — we are confronted in the cases described here with a partial form of retardation: in the non-premature cases we have a full-term child with a central nervous system which has not reached full term.

The question of an anatomical concept of this 'unripeness' will have to be answered by microscopic investigations.

Our present knowledge on this point is certainly very far from complete; but it seems as though there are already indications as to the position and nature of this retardation phenomenon. Here I refer to the fact, already mentioned, of backwardness in development of the myelin sheaths in some autopsied prematures.

From this I venture to deduce that the nerve elements encased in these sheaths have also not reached full maturity. It even appears that a choice must be made at this point, but there is not yet sufficient basis for such a choice, i.e. for a decision as to whether such a nerve element was primarily at fault or whether the conditions for satisfactory myelin formation were lacking and this had its repercussions in a lagging behind of functions.

Since all growth is subject to hormonal influences, it seems possible (especially in cases with a pre-existing deficiency) that the sudden interruption of the supply of maternal hormones at birth may have an unfavourable effect on the child's hormonal equilibrium, and hence also on the growth of the brain. The occurrence of disturbances will be promoted bij any pre-existing, primary deficiency of hormone production in the foetal organism, and also by toxic factors which may act either locally or via the endocrine system.

The experiments of Stephen Zamenhof and Arthur Weil on the effect of hormonal factors on growth of the brain in tadpoles and albino rats might provide support for such a hypothesis and also, perhaps, a possibility of rational therapy in the distant future.

In man also, examples are known of the influence of disturbances in the maternal or foetal endocrine system on the growth of the brain, and of defects and retardations ascribed thereto (Ida Mann).

Physiologically, such factors are manifested first in an absence (often total) of development of certain mental, motor and sensomotor functions and later in a retarded, very gradual development thereof. The popular term 'backwardness' seems to me to represent the concept very correctly, since it gives the idea of a rate and leaves the possibility of 'catching up' entirely open.

Many of our little patients do really catch up on their arrears in the long run; only a few of them are left with residual defects in the form of sensory (visual) or motor disturbances.

The occurrence of motor defects of the extremities with the secondary consequences thereof (contractures) might also be connected with maturation and retarded myelination of the nerves concerned, as a primary, hereditary factor. Conditions such as metatarsus (pes) adductus congenitus or metatarsus varus congenitus would then be of secondary development, as a result of temporary or permanent predominance of certain muscle groups, while the same might apply to hand of finger deformities and perhaps also to Little's disease.

Upon surveying the cases described above we find that the development of these children, regarded from the general and more particularly from the ophthalmological point of view, has much in common with that of our other strabismus patients. In fact, apart from the initial blindness of the former group, the course of events is entirely parallel in the two groups. Further consideration will be given below to this (really prenatal) stage in the development of sight and the consequent reactions in these children, and to the gradual transition to the condition found in the normal baby at birth.

As regards the time of birth, 4 of our 12 patients were prematurely born; 2 were 7-month babies and the other 2 were born at $8\frac{1}{2}$ months. In all cases the actual birth had been normal and the pregnancy undisturbed.

Ten of these children were examined for the first time between 2 and 6 months after birth; the other two were 3 and 4 years old respectively.

The anamnesis was practically identical in all cases. All had originally given the impression of blindness, to parents and others; this condition sometimes lasted for months and led to the consultation of an oculist. In all cases the eyes during this period performed restless wandering movements, largely dissociated. The pupils were sometimes wide and without any reaction; in most cases, however, they already showed some reaction (although sluggish) to light at the time of the first examination. As a rule the fundus showed little or no pigmentation and in 6 cases the papilla was more or less grey.

As set out in detail in the foregoing, this first stage gradually passes into a second in which the pupil reaction becomes more distinct and the eyes begin to show signs of light perception. The movements become more associated and faster and acquire a nystagmoid character. At first hesitatingly and then more and more decidedly, fixation and following movements to light and to presented objects make their appearance and the child begins to make grasping movements. It now begins to smile and the face, originally blank and mask-like, gradually becomes more expressive.

The nystagmoid type of movement — with which step-wise and jerky movements in all directions may also be intermingled — now gradually changes to a quieter form more like ordinary nystagmus, while the amplitude becomes smaller and the movements slower. Conjugate movements, evoked by optical stimuli, now get the upper hand and from time to time the eyes remain still for a moment, as though fixing. This stage then merges gradually into the last stage, in which the eyes come to rest completely and a condition is now reached which corresponds to that habitually seen in normal babies at birth.

Our knowledge does not yet suffice to provide an answer to the question: how are these movements produced? At the most, a few conjectures may be ventured.

So long as we are still in ignorance of the detailed anatomical and histological picture of the connections between the eye and the various parts of the brain which have reached completion at any given moment, we shall be obliged to make use of the signposts provided by comparative physiology and to search among clinical data for possible signs of relationship.

In our discussion of the several stages of ocular movement in the infant we pointed to a certain analogy that seems to exist between the types of ocular movement observed on ascent through the animal kingdom and those seen in our patients at a certain stage of their development.

In this connection we sought support in the ideas of Hughlings Jackson, Von Monakow and Gourevitch and in the comparative embryological, histological and physiological studies of Minkowski. All these investigators are convinced that the ontogenetic development of motoricity presents a repetition of the phylogenetic. Minkowski writes of these researches: 'they permit us, as it were, to follow or at least to reconstruct the order of appearance and development of the nervous functions (movements, sensitivity, reflexes etc.), starting from their early stages and following their progressive, embryonal and foetal transformation parallel to definite anatomical and functional changes in the nervous system.'

He describes in turn the first, idio-muscular or aneural movements in fishes; the first swimming movements in amphibians and the primitive movements seen from the 4th week onwards in mammals. These last-mentioned movements are at one moment bilaterally symmetrical and at another alternating; a few weeks later they again become asymmetric, arrhythmic and highly variable, while here also the movements of the anterior extremities still predominate. In the first stages they are weak, diffuse, very variable and little co-ordinated. Later they become stronger, more regular and better co-ordinated.

Turning to man, Minkowski then studied movements of 80 human products of conception at various stages of development. Bergeron writes of this investigation: 'By directly observing the movements and reflexes of the still living foetus, by comparing these facts of physiological observation with the morphological development — in particular that of the nervous system — and by viewing these data in the light of general biological and embryological concepts, we may, according to Minkowski, distinguish the following stages in the functional development (especially nervous) of the human foetus:

Stage 1: (5—8 weeks or 1.5—4 cm.). The movements of trunk and extremities are slow and probably idio-muscular.

Stage 2: (about 2 months or 4—5 cm.). The movements are slow, often vermicular, asymmetric or arrhythmic and extremely variable. While the extreme slowness and the vermicular nature of the movement might point to excitability of the muscle itself, the variability and irregularity might suggest the first, primitive activities of the central nervous system.

Stage 3. (3—4 months or 5—15 cm.). The movements now become more intensive, faster, even abrupt and sometimes choreatic; spinal reflexes are present from the 3rd. month. At the end of the 4th. month the first myelin sheats appear in the anterior roots of the lower cervical part of the spinal cord, at the level of the centres for innervation of the neck and anterior extremities (cervical reflexes!); this is soon followed by the appearance of myelin sheats in the corresponding posterior roots.

Stage 4: (4—6 months or 15—30 cm.). The movements are now faster and better characterized, this being accompanied by a restriction of the reflexogenic zones and a better canalization and co-ordination of the various reflexes (foot-sole, neck). Myelination advances cephalad and reaches the nuclei of the motor cranial nerves: first the vestibular and about 2 months later the acoustic nerve. The first myelinated axons now appear also in the tegmentum, medulla oblongata and pons.

Stage 5: (6—9 months). The reflexes are now better regulated and some new reactions, evoked via the sense organs, are added to those already present. This stage is characterized by the further progress of myelination of certain tracts and fibre systems in the medulla oblongata, cerebellum and mesencephalon (including nuclei and roots of the VIth., IIIrd. and IVth. cranial nerves and the lamina quadrigemina).

In the diencephalon the ventral thalamic nuclei, the nerves, the optic chiasm and the optic tracts are gradually included in the process of myelination. Outside the primordial regions, the hemispheres at birth show only the very earliest signs of myelination, in the frontal and parietal regions. This is, thus, the subcortical stage, which is characterized as regards motoricity, at the time of birth, by a predominance of the tegmental, mesencephalic and palleo-cerebellar reactions.

The postnatal period of 6—8 weeks that now follows is characterized by the growing influence of the motor cortex on the subcortical and spinal reflexes, via the pyramidal tract which had started its myelination just before birth. This also accounts for the fact that the foot-sole (plantar) reflex does not as a rule acquire its plantar character until the 3rd. month.'

Bergeron, who studied the movements of infants in the first 3 months and from whose work we have gathered much of the information presented above, considers his findings to confirm those of Minkowski. He also sees in the phylogenetic development of motor functions the basic pattern of the development of motoricity in the human infant.

If we now adopt the view — practically universally accepted — that the development of motor functions is correlated with the development and the stage of development of the central nervous system, we then explain the fact that a given reaction is still absent, faulty or insufficient as being due to the still incomplete building -up or maturation of the central parts, their pathways and their centripetal, centrifugal or mutual connections. This being accepted, there seems to be no reason why the same arguments should not apply to the movements of the eye and to the optomotor reactions in particular.

From the point of view of physiology, it is incorrect to ascribe disturbances of these reactions in newborns or young babies to other than physiological factors. Starting from this conviction, I came to the conclusion that squint can be and must be explained by factors of this kind.

Returning to the apparently abnormal eye movements of our patients, we note that the movement-pattern seen at the various stages showed — both in nature and in chronological order — a high degree of resemblance to that seen in the successive stages of ontogenetic development of the general motoricity, so that in a certain sense we may speak here of a repetition.

An explanation of these peculiar movements will, in my opinion, require to ascribe an important rôle to evolutionary influences.

As regards origin of the stimuli, we may assume with virtual certainty that in the stage of effective blindness the movements are produced by idio-muscular and vestibular stimuli and by those from the proprio- and enteroceptive spheres. The wide extent of the reflexogenic zones and the pronounced radiation of stimuli are important points in this connection.

Here we may take some data on reactions probably evoked

by optical stimuli, as observed clinically in analogous conditions. A drawback is that most or all of these relate to observations on adults. Many authors (Potzl, A. Fuchs, Economo and others) have published observations on the restoration of sight after bilateral cortical lesions of the occipital pole, in which, in addition to a vague 'organ sensation' (more elementary than light perception), a trace of fixationmovements was detectable as a first sign of the return of sight. These observations may perhaps mean that we have to regard the primitive fixation movement as the first sign of an awakening of the cortical function. In the cases reported this sensation was still so elementary that the perception of light and darkness (without any further differentation) did not become possible until recovery had advanced further and dark-adaptation had appeared.

Kestenbaum (1930) found that in some cases of homonymous hemianopsia, with a lesion of the middle or posterior part of the optic radiation, the normal gliding following movement was replaced by a succession of jumps (cogwheel movements).

On these grounds we assume that cogwheel movements in the newborn are signs of a primitive stage of development of the fixation reflex. The results of investigation of the following movement in babies up to 4 months (Kestenbaum, Bing) point in the same direction.

The rare cases, reported by Bielschowsky and Holmes, of bilateral lesions of the corticotectal (afferent) fibres in the occipital cortex present an extensive disturbance of fixation. As in nystagmus, both eyes here wavered continually to and fro in attempts at fixation. Fixation, accommodation and convergence were disturbed.

This may perhaps suggest that the nystagmoid movements and the subsequent nystagmus are connected with disturbances in the corticotectal paths. In the infant, the stage of development of the macula will also possibly play a part.

Unilateral lesions of the corticotectal path also lead to fixation difficulties, especially on looking sideways. The fixation cannot be maintained and the eyes return to their position of rest each time.

With unilateral lesions, thus, jerky nystagmoid movements similar to those seen in our cases are observed.

If we now remember that the autonomous rhythm of these various movements can be crossed by reactions evoked along the above-mentioned subcortical paths; that the reflex disorder (although often bilateral) may be unilateral; that every displacement of the eye from its original position has in it the nucleus of an opposing movement to restore the disturbed equilibrium, and if we consider the disintegrating influence of an absent or imperfect connection to the cerebral cortex (Le Roy Conel, III, p. 146) and possibly of the latter with the striatal system (Lesné et Richet, Kurz) — a disintegrating influence that is perhaps due partially to insufficient inhibition of stimuli (McCulloch) —, we shall be able to form some idea at least of the manner in which this complex system of primitive reactions is produced. In this way we may perhaps ultimately be able to find an explanation. The whole question is so complicated that it is not surprising that little or no attempts have so far been made to analyse and explain this system of movements.

Such analysis would, however, be worth the trouble, in view of the valuable information it might provide with regard to the stage of morphological development, consolidation and maturation reached by the central nervous system. Moreover, clinical and physiological research would in this way stimulate anatomical research and provide signposts for it.

The considerations advanced above bring forward once again the question as to the cause of these serious and complicated phenomena. An important pointer is here the fact that in most cases all these seemingly alarming manifestations disappear spontaneously in the course of further development, usually in a few months. From this favourable development of the affection, as seen in my cases and those of Beauvieux, the only logical conclusion seems to be that all the phenomena observed were the result of a disturbance or retardation of the normal process of growth.

In connection with our knowledge of the myelination of the central nervous system in general, for which we are indebted to the investigations of Flechsig, Déjerine, Minkowski, Le Roy Conel and others, and that of the optic tracts in particular (Von Hippel, Westphal, Pfeiffer and Beauvieux) it seems obvious that we should think in the first place of a retardation of myelination [1]) as cause of the observed phenomena, the more so as the building-up of the tracts and white matter is still actively in progress at the time of birth. It goes without saying that the anatomical growth (de Crinis) and physiological

---

[1]) See p. 111.

maturation of the ganglion cells, which are normally correlated with the myelination, [1]) will at the same time also suffer a retardation.

It is this retardation of the general consolidation and growth of certain parts of the central nervous system that — as mentioned before — I wish to designate *myelogenesis retardata;* this leads to a disturbance in the conduction of impulses and hence also in the functional consolidation and may — according to the situation and extent of the affected region — become manifest in very different ways and in various forms.

We are still left with the question of the feature in common between the group of cases described above and the cases of strabismus in 'normal' children.

As can be seen from a comparison of the results of examination of squinting children (see Chapter III) with the observations recorded in the case-histories in this chapter, *the characteristic feature in both groups is the appearance at a certain stage of development of the same disturbances in the genesis and building up of the optomotor reflexes,* although these disturbances may differ in degree.

The only thing that is absent in the case of the squinting child is the period of severe disturbances of motoricity, as a result of the action of optical and other stimuli, which in the other group occurred at a time when the establishment of the central connections was still very far from complete, so that it was inevitable that both the medium and the tension-pattern should differ considerably from normal. This comparative study shows that both groups present the same deviation from the chronological order in which the optomotor reflexes normally become manifest. In both this gives rise, as the first reaction, to a predominance of the adduction reflexes over the reflexes to conjugate movements and the abduction reflexes.

In both this predominance of the adduction leads to the same consequences with respect to the position of the eyes; *in both groups the result is that a convergent squint develops.*

In both groups also, the further course and the tendency to spontaneous correction of the affection as a whole (retardation) i.e. both of the faulty position of the eyes and of other factors, is exactly the same while further, both these groups of children

---

[1]) Cf. Le Roy Conel. III.

show great similarity in very many other points connected with the development, general nervous condition and behaviour.

It goes without saying that only morbid anatomical findings can provide objective proof that the same cause of the disturbances is really present in both groups. A painstaking search for such proofs is being carried on. Until these can be produced, however, we shall have to be satisfied with the results of comparative physiological research and clinical analogies. The EEG (electro-encephalo-gram) might provide objective support here.

On the grounds of our data and our present knowledge of the myelination process we may consider ourselves justified in concluding that the affection is produced by the same cause in both groups and that the cause in all probability lies in a retarded myelination [1]). The disturbances observed would then be the manifestation of this retardation phenomenon.

These considerations have emboldened us to place the title 'myelogenesis retardata' at the head of this chapter.

Thus, everything connected with the occurrence of squint will depend on the condition of the cerebrum — and in particular the occipital lobes with their associated tracts and connections — at the time of birth. This applies especially to the reciprocal connections between the terminations of the crossed and uncrossed fibre bundles and the fibre systems which connect these with the motor areas.

If at this moment the pathways — also in the afferent (corticopetal) system — are still far from complete, the child will be born blind; if there are slight disturbances in the connections and the efferent (corticofugal) fibres and tracts, there is a very great chance that the infant will begin to squint soon after birth. If however, all tracts and connections are completely myelinated at birth, the possibility of the occurrence of a primary squint can be regarded as excluded. Since myelination occurs partly under the influence of function, the last-mentioned case will presumably never actually occur. We may therefore assume that practically every baby is born with the potentiality of squint. Whether it will actually squint depends (apart from secondary causes) on the possibility of a timely establishment of the binocular junction.

---

[1]) See. p. 111.

CHAPTER VII

THE SIGHT OF INFANTS

U PON observation of the way in which a baby a few days
old looks around it in an apparently normal way, with the
eyes wide open, even an ophthalmologist will hardly be
inclined to doubt that the child does really see, i.e. receives and
analyses visual impressions. This seems to be a matter of course.

The ophthalmologist finds himself in eminent company in
this opinion: such authorities as Donders and Engelmann
observed 'binocular fixation with alteration of convergence in
a male infant barely an hour after birth' and even saw a child
that 'a few minutes after birth quite definitely fixed binocularly
an object held in front of it and not only followed lateral
movements of this but also increased the convergence when the
object was brought nearer and decreased it when this was moved
away'. Hering also noted 'convergence and relative or weak
absolute divergence movements' in newborns.

There is no doubt as to the correctness of these observations;
only the possibility of a different interpretation remains open.
Can a baby, when the optical stimuli have been able to exert
their action on the previously virgin soil only for a few minutes
or even an hour, really be capable of fixation and faultless convergence and thus already have reached, at that moment, these
two peaks of the curve of the phylogenetic development of vision?
It is not possible to give a direct answer to this question. An
approximate answer can be found, however, by ascertaining
whether the required anatomical substratum is already present
in its completed form and whether the various connections and
optomotor reactions needed for these higher functions have already developed.

That these are still very far from complete at birth has been
shown by the investigations of pre- and postnatal development
of the brain carried out by the workers mentioned in the previous
chapter. In this connection I should like to draw particular at-

tention to the work of Flechsig and to the recent extensive and admirable work of Le Roy Conel who, during the last war, gave a description of the postnatal development of the brain in the 1st., 3rd. and 6th. month of life, also on comparative lines.

A glance at the three highly instructive photos of preparations made by Flechsig, which are reproduced below, will suffice to convince the reader on this point. Even a simple comparison of the volume of the cortex in relation to the rest is enough to show what arrears have to be made up in the first 6 months.

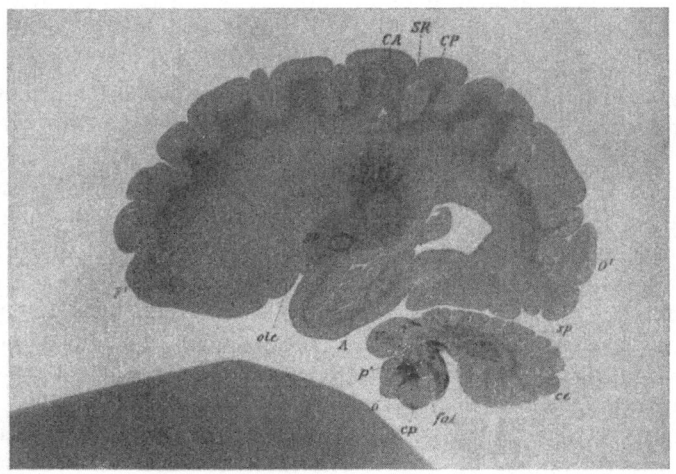

Cerebrum of a premature 8 months. Myelinated fibres black. (from: Anatomie des Menschlichen Gehirns und Rücken-marks auf Myelogenetischer Grundlage. Paul Flechsig Bd. I Leipzig 1920).

There is no question here of any proliferation of ganglion cells — this came to an end about the 5th. month of intrauterine life — but only of an increase in size and structural development of the cells and, in particular, of the building up of the white matter. After the 1st. year this building up proceeds more slowly. It can be reckoned that the connections of the peripheral organ with the occipital cortex and the fibre system which belongs thereto have been more or less completed by about the end of the 2nd. year. It is held, however, that about 10 years more are required for the brain as a whole to reach its full

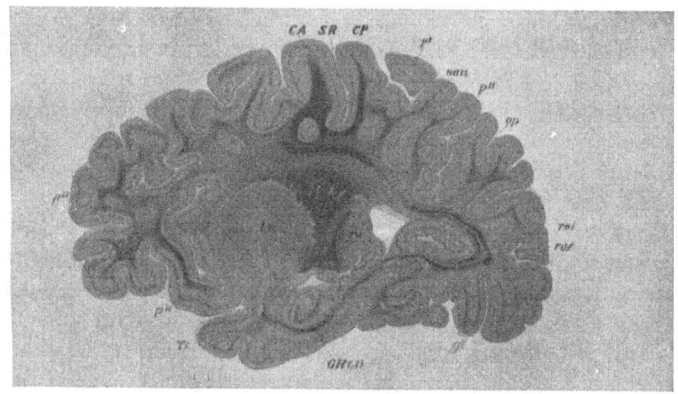

Cerebrum of a child aged 7 weeks.
Myelinated fibres black.
(from: Anatomie des Menschlichen Gehirns und Rücken-
marks auf Myelogenetischer Grundlage Bd. I Leipzig
1920 by Paul Flechsig).

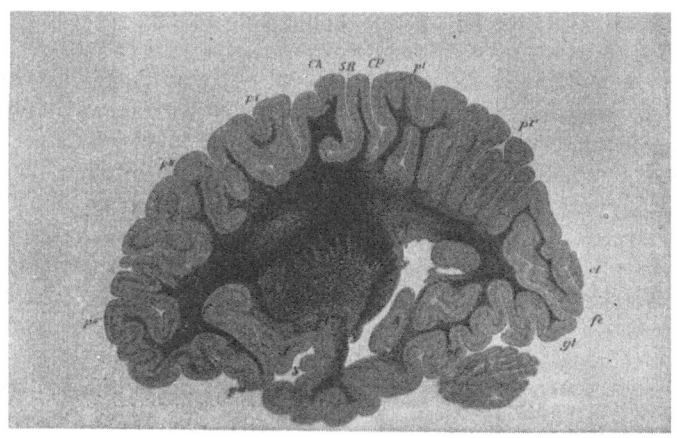

Cerebrum of a child aged 4 months.
Myelinated fibres black.
(from: Flechsig Anatomie des Menschlichen Gehirns
und Rückenmarks auf Myelogenetischer Grundlage.
Bd. I). Leipzig 1920.

development. Light stimuli and exercise will have a stimulant and constructive effect here. Berger and de Crinis have shown the cytodendrogenesis also runs parallel to the function.

Since function can only proceed hand-in-hand with structural development and myelination, it will be obvious that a considerable period of time will be required for the attainment of the highest, in many cases superposed functions.

If we now, after having acquainted ourselves with this general anatomical picture, turn our attention to the finer structure of the brain, we see — as we have already remarked — that the further growth of the ganglion cells still has to take place after birth. In this connection it should be remembered that the maturation of the cells in the area striata takes place first and that the development of the ganglion cells in the para- and peristriate areas follows step-by-step after this has been completed. This process takes years for its completion and is correlated with the development of the mnemic-associative and memorial functions (de Crinis).

The starting point of the motor impulses evoked by optical stimuli is believed by Economo to be located in the large pyramidal cells of the limes parastriata-giganto-pyramidalis.

Turning now to the white matter, the finer fibrous structure of which is related to the function of the region in question — the area striata representing a type that is not found anywhere else, — we are struck in the first place by the fact that the myelin sheath surrounding the fibres is still very incomplete at birth and in some places has not even started to appear. Since it is generally held that the process of conduction of stimuli in the fibre is dependent on the intactness of its myelin sheath, it must be concluded that the connections of various parts of the brain, both among themselves and with the periphery, will show serious imperfections at many points.

A general survey shows us that the myelination of the cortical fields of the sensory spheres is complete at birth. Proceeding from the subcortical centres, the myelination of the corticopetally-conducting axons has made an early start. The fibres which emerge first from the subcortical centres are those which have to make contact with cortical fields that are connected with the peripheral sense organs or their nerves (corticopetal tracts), e.g. the tract: corpus geniculatum laterale — radiatio optica — area striata.

The myelination of the centrifugal tracts at birth must not, however, be assumed to be absolutely complete. Flechsig says on this point: "The primary optic radiation is (at birth K.) myelinated throughout its extent; about in the middle, between the dorsal and ventral sides, it is adjoined by some myelinated fibres which are arranged in distinct bundles; these form part of the secondary optic radiation: primary system of the secondary optic radiation. Dorsal and ventral to this the secondary optic radiation (the inner sagittal layer) is still unmyelinated (p. 23)." "In actual fact a large number of important fibre bundles of the corona radiata are still unmyelinated in post-mature new-borns. These are:

(1) a considerable part of the secondary optic radiation, (Flechsig) lower portion;

(2) the posterior superior colliculus;

(3) the anterior colliculus (middle part).

Most of these fibre tracts do not acquire continuous myelin sheaths until between the 2nd. and 4th. month of life. They are not, however, as one might think upon superficial consideration, connected to future cortical fields but are all, notwithstanding their later development, connected to primordial regions" (p. 15).

"In view of the fact that birth may be delayed a full month over term, it is not impossible that here and there a considerable number of myelinated fibres may appear after a shorter period of extrauterine life" (p. 15).

"Towards the end of the 4th. month after birth all the distinguishable fibre tracts are myelinated as regards their main fibres. It is true that many other fibres, particularly in the cortex, still have to undergo myelination, but these are either collaterals or more scattered nerve fibres which do not belong to large bundles. With these exceptions, thus, the process of myelination of the larger fibre tracts may be regarded as completed at about 4 months after birth at term." (p. 10).

Flechsig rightly regards his own findings which in my opinion have not yet received the attention they deserve — as very important. His remark that "The process of successive myelination provides a programme for anatomical research — a programme that is followed by Nature herself — leading to profound conclusions as to the process of development and the chronological order of appearance of functions, especially mental" (p. 8), deserves full endorsement.

If we now place the anatomically-preformed possibilities of the central organ, with respect to conduction, to transformation, synthesis and analysis and to motor reaction, alongside those of the peripheral organ, we see that on anatomical grounds alone there is more than sufficient reason to doubt whether a baby can 'see'.

Examination of this peripheral organ shows that, as regards the image-forming apparatus, the optical axis cuts the retina in a point lying between the fovea centralis and the papilla. From the fact that the fovea lies temporally and the papilla nasally to the optical axis it appears that the medial half of the eyes is slightly less well developed than the lateral half — at any rate if we accept the position of the optical axis as marking the division between medial and lateral.

The cornea is more sharply curved and the radius of curvature of the anterior and posterior faces of the lens, which is almost spherical, is also much smaller than in the adult condition.

Of the ciliary muscle, the circular part still has to undergo its whole development after birth. This postnatal development seems to proceed in correlation with the refraction of the eye (Ida Mann). The length of the optical axis is about 2/3 of that in the adult.

As regards the receptive apparatus, the peripheral retina has practically completed its differentiation at the time of birth. The macula, however, will need a further 6 months of growth before it can be regarded as capable of full function. In many cases, however, the pigment epithelium seems to be still only moderately developed at that moment.

In connection with the conduction of stimuli, the following may be said about the optic nerve. At birth the myelination of the optic fibres has barely reached the lamina cribrosa and a further period of about 10 days will be required for completion (Ida Mann). Both the image-forming and the receiving and conducting systems, thus, are still in the process of being built up.

On the grounds of this anatomical survey we may state — and this is very important — that in the normal baby at the time of birth there are still numerous gaps in the myelination of the optical tracts and connections, which are likely to cause disturbances of conduction, and that the peripheral organ is also still at an unfinished stage.

From an anatomical point of view, thus, there will theoretically be the following 3 possibilities with respect to light perception and motor reactions. The infant enters the world:

(a) with disturbances of conduction in the corticopetal tracts, which can be further divided into disturbances affecting the first or the second neuron;

(b) with disturbances in the corticofugal tracts or perhaps in the conduction of stimuli along both corticopetal and corticofugal tracts and connections, or

(c) without disturbance of conduction.

In the cases as under (a), with conduction disturbances in both neurons, neither pupil reaction nor light perception will necessarily be present. With disturbances affecting only the second neuron the pupil reaction to light may be present and it is perhaps permissible to assume that there is a vague, subcortically-determined sensation of light.

In cases as under (b) with disturbances in the corticofugal tracts only, there will be perception of light and a pupil reaction to light, with absent or highly imperfect optomotor reactions.

In cases as under (c) both the light perception and the motor reactions evoked by the light stimulus may all be normally present.

All possible transitions between these 3 stages are conceivable, while the possibility of unilateral disturbances must also be taken into consideration. Finally, small defects and retardations may alter the chronological order in which the various reflexes normally develop, or lead to an over-all delay in the manifestation of these reactions in their normal order. Here one thinks first of the connection between optomotor reactions and squint.

In this connection I should like to emphasize once more that my hypothesis as to the cause of these disturbances in strabismus does not involve the introduction of a new disease factor but that we are concerned here exclusively with a certain degree of accentuation — attracting attention on account of its consequences — of an otherwise perfectly normal physiological process that occurs during the development of every child.

In view of its frequency of occurrence, squint as a consequence of this process appears to lie just at the limit of normality.

Children with retarded myelogenesis show an exacerbation of the process in question that brings it just outside this limit, while those with permanent disturbances, possibly of toxic origin, come into the pathological category.

From the previously described cases of retarded myelogenesis and the results of my investigations on normal babies, described below, it appears that practically all the theoretically conceived cases do actually occur in reality.

This in itself is equivalent to stating that in a number of cases the development of optomotor reactions will be disturbed.

In order to gain some insight into the optic perception and the development of the motor reactions correlated with it, I made a study of the optical reactions of 87 newborns in the Groningen Obstetrical Clinic during the first few days of their lives and followed them up for some time afterwards. In 2 of these infants the optical reactions were examined before any optical stimulus had been able to act on the retina. This was achieved by carefully covering the eyes with a well-fitting damp bandage as soon as the head was born and taking the child, immediately on completion of its birth, to the dark-room for examination.

The movement of the eyes were now followed with the aid of a piece of the membrane of a raw hen's egg, upon which a small amount of a fluorescent compound had been placed. This membrane adhered easily to the cornea and the movements could be followed by the use of ultraviolet light. The eyelids were kept open activily or sometimes also spontaneously. It was now found that both eyes executed slow, wandering movements, in all directions quite independently of each other. When the ultraviolet light was replaced by photographic red light the movements continued in exactly the same way.

When the illumination was changed to a 40 watt electric lamp at a distance of about 1 m., with incidence on both eyes at once, it was found possible to evoke in one of these babies a conjugate fixationmovement of very short duration, after which the eyes immediately resumed their wandering movements. In both infants these movements were found at first to continue also in ordinary daylight. The second infant did not react to the artifical light with a fixationmovement.

In both children, turning of the head to the right or left was followed by a distinct compensatory reaction; with movement of the head in a sagittal plane this was less marked.

The cervical reflexes were not studied.

On the grounds of these investigations it may be assumed that the slow, monocular wandering movements described above are already present in the foetus in utero, from a given point in development onwards, possibly as a reaction to idio-muscular stimuli.

All conceivable positions of the eyes, including of course the convergent and divergent squinting positions, were seen in these children during the investigation in question. Further study showed us that these dissociated wandering movements can be observed in all babies for a certain length of time after birth.

They continue for periods which vary individually from a few weeks to some months. After 1 to 3 weeks they are seen to decrease in intensity in most cases while they also become faster, after which they gradually disappear or appear only inter-mittently, also in the short waking periods in which the eyes are spontaneously kept open. The impression is gained that the general condition and the increasing influence of the light stimulus are of importance here.

During sleep they are still observable up to 6 months or longer in many cases, and also at the moments of falling asleep and waking.

The 'squinting of babies', which is imitated by these wandering movements, may thus be encountered during this period. In this form it has nothing in common with true squinting (evoked by optical stimuli) and is in fact nothing more than a demon-stration of the dissociation of the ocular movements.

The examination of the optomotor reactions of the other 85 infants showed these to differ widely, both individually and in connection with age, so that the results can best be reported in groups.

A pupil reaction to light was observed in all babies quite soon after birth. In many cases this reaction in the first few days was very slow and sometimes vermicular. It is not known in how far this may be ascribed to a direct action of light.

A fairly large group did not react with an eye movement to a moderately strong light stimulus (9 V. lamp) at a short distance from the eye. A smaller group sometimes showed a transitory (conjugate) fixationmovement in response to this stimulus, a very short time after birth. With increasing age (3 weeks onwards) the number in this group increased rapidly.

Only a very small group showed in the first week not only a fixationmovement in response to light but also a following movement or the beginning of one.

A not inconsiderable group of babies showed spontaneous monocular reactions, while jerky and nystagmoid movements were seen in a rather smaller group. Occasionally the observed picture was strongly reminiscent of ordinary horizontal nystagmus. The extent to which we were here confronted with reactions to optical stimuli could not be ascertained.

Only in one baby — and I do not consider this a normal case — was I able to observe at the end of the first week a bilateral adduction movement in response to a source of light held a short distance in front of the eyes in the median line.

With most of these children I was, of course, only able to continue the examinations for one to a few weeks. For further investigations I had to take children in the local hospitals. Here I found that the compensatory (labyrinthine) reactions remained practically unchanged. Further, in the overwhelming majority of children examined it was possible to elicit a conjugate reaction to light after only a few weeks; this was achieved more readily from the nasal half of the retina. [1]) Monocular reactions to stimulation of the nasal or temporal half of the retina were not shown by these (normal) children during the first months. A bilateral adduction reaction ("convergence reaction") could not, as a rule, be evoked until after the 3 rd. month. This also applied to optokinetic nystagmus.

Summing up, we may thus say that during the first few days and sometimes weeks of their lives a number of these children did not show any motor reaction to light stimuli, except for the pupil reaction. In all probability, therefore, the connection to the optical cortex or that of the cortex with peripheral gaze centres was still unformed or only very imperfectly formed.

In a small group, in some cases even from the first day, a fixationmovement was observed in response to the light of the electric ophthalmoscope held at a short distance from both eyes, although the ability to execute followingmovements with this intensity of stimulus was not present.

For this group it must be assumed that the connection of the peripheral organ to the area striata had been established and that the corticofugal tracts were also functioning, although still incompletely.

Only a few children showed, in addition to fixationmovements, followingmovements at a rather early age, although these movements could only be kept up for a moment, showing

---

[1]) The conjugate movement seems to be a primitive reaction. (Rademaker and Ter Braak.)

that the reflexes elicited from the more central part of the retina were gradually coming into function. Kestenbaum found these reflexes only after the 3rd. month in most cases. My experience is that they may occasionally be seen earlier.

It thus appears that some children are probable practically blind at birth; others show signs of some degree of light perception and a 3rd. group is already capable of motor reactions to optical stimuli.

To evoke a (conjugate) fixationmovement it is generally necessary at first to allow the light stimulus to act on both eyes. Gradually it becomes possible to achieve this also by stimulating only one eye, the nasal half of the retina giving a readier response.

I never observed a unilateral or bilateral adduction movement (in my opinion it is here not yet justifiable to speak of a convergence reaction) in normal babies before the 4th. month.

The blinking reflex to menace does not as a rule appear before the end of the 3rd. month either (Bergeron); the first smile is generally seen after 6 weeks. The condition of the image-forming apparatus and the anatomical development of the ciliary muscle and the macula are such that there can be no question of sharp images in the first few months, while the absence of convergence makes binocular seeing of near objects an impossibility, apart from the fact that the fusion reflexes are certainly not further than the very earliest stage of their development Vision is, however, a cerebral and not a peripheral function.

By regarding vision as a function of the peripheral organ and, by the same token, squint as a peripheral disturbance, while neglecting their central background, we have created imaginary problems which have led to many misconceptions, as Zeeman has pointed out in another connection.

And what is now the condition of this cerebral function in the infant? To start with, it is not quite correct to speak of 'the sight', since there are individually differing stages not only in the stage of primary development but also in the adult.

We may conceive of vision as being the result of the co-operation of 3 components: a perceptive, an optomotor and an associative. Only the faultless co-operation of these 3 factors can give us, as a conscious sensation, an impression of objects belonging to the outside world.

The inclusion of the peri- and parastriatal areas in the process, and hence of the connections with other parts of the brain, is very gradual, taking years to complete (de Crinis). It is therefore obvious that, after the development of the optomotor component, a very long time will be needed for the building up of the associative connections necessary for the recognition and interpretation of objects, and that in the young infant these qualities will still be very little developed.

It may thus be deduced that in the first few months there will only be elementary perception without the possibility of analysis, so that we may at the most only speak of what Parsons calls protopathic sight.

The investigations of Wald and Burian have shown what qualities may, at the highest estimate, be attributed to this protopathic sight.

These workers ascertained objectively that gradations exist in human vision. Their study of the amblyopic eye showed that the threshold value of this for light and colours, both after light and after dark adaptation and measured centrally and peripherally, was completely normal and entirely in agreement with that of the fixing eye. In the power of fixing and localizing a point of light peripherally or centrally the amblyopic eye also showed practically no difference from the good one.

Since the visual acuity of the amblyopic eye in their cases was never higher than 20/200, Wald and Burian concluded on the grounds of the otherwhise normal functions of this eye that the possibility of form-perception must to some degree be separate from that of simple light-perception and space-projection.

They also pointed to the great measure of correspondence existing between the vision of mammals deprived of the occipital cortex and that of the amblyopic eye. Such animals entirely lose the power to distinguish shapes and objects, while their powers of reacting to light stimuli or distinguishing brightnesses are only slightly decreased.[1] The power of localizing stimuli in their field of vision is also preserved to a sufficient degree.

On these grounds they conclude that in mammals in general the anatomical substratum for light-perception is partly sub-

---

[1] Ten Braak noted that in order to evoke optokinetic nystagmus in decerebrated dogs and cats the objects had to be moved much more slowly than normal.

cortical and that a certain degree of visual orientation in space
is subcortically established, whereas the seeing of shapes and
objects requires the participation of the cortex.

Although human vision seems to be bound in all respects to
the cortex, the existence of such a complete functional separation
between the different qualities of vision in the amblyopic eye
as the findings of Wald and Burian indicate compels us to
assume that these functions are anatomically separated to a
certain degree. These authors consider that in man also the perception of form is a higher and the perception of light and
spatial localization a lower cortical function. [1])

I should like to add here a few personal observations on
children with myelination disturbances in the stage of incipient
light-perception.

Here it is possible to observe clearly the struggle between
the subcortical and the cortical centres for the hegemony over
the position and movements of the eyes. On the one hand the
eye movements are still entirely subcortical in character, as can
be seen from the nature and type of the movements, while on
the other hand signs pointing to a vague light-reaction are
already present. It is tempting to imagine, on phylogenetic
grounds, that this primitive perception of light — which is
certainly still very diffuse at this time — might also be subcortical.

While in these cases it remained uncertain whether the light
perception was subcortical a rather greater degree of probability
was available in the case, described as No. 8 in the preceding
chapter, of a child with extensive disturbances of myelination.

This child showed distinct cervical reflexes; when it was laid
on its back and the head was turned about 90° to the R. or L.
there was an extension movement of the homolateral leg, while
the other was drawn up in the flexed position and kept there
as long as the position of the head was maintained. On account
of intercurrent movements the reaction of the arms was not
clear. I was also unable to observe any sign of reaction of the
eyes.

If it is true that the presence of cervical reflexes shows that
the cerebrum is to a great extent excluded, as would appear from
the investigations of Brouwer, Stenvers and others (Magnus:
Körperstellung, pp. 126 and 127), the optic reactions noted in
this case might point to a subcortical process and be comparable

[1]) In this connection see also Chap. VIII. Amblyopia.

with the reactions observed in higher mammals and apes after removal of the occipital cortex. In addition to the first, sub-cortically-regulated eye movements, the first 'seeing' of many newborns might well be subcortical, while, conversely, the optical stimuli in question here might lead to eye movements without participation of the cortex — a problem which has recently occupied the attention of various investigators (Rieken, Best, Winkelman). The perception and optomotor reactions then presumably become gradually transferred to the cortex as the child's development proceeds.

Professor Brouwer, to whom I submitted this problem in private correspondence, expressed his ability to accept this view as a possibility.

Recent research and the much wider concepts of the activity of and representations in the cerebral cortex which are now accepted have also induced Duke-Elder to reason along these lines. He writes:

'Indeed we may go further and say that mental activities are not localized in the cortex but are probably associated also with subcortical regions.'

'It is probable that the migration of activities seen in phy-logenetic development from the stage wherein conciousness resided in and behaviour was determined by the upper portion of the mid-brain and the cerebal cortex was merely an olfactory receptor, has not been complete, and that certain aspects of mental activity-and indeed of visual activity-especially those aspects endowed with emotional tone and associated with the most primitive responses, remain closely related to the vegetative life, which is mainly integrated in the thalamus.' (see Holmes: Introd. to Clin. Neurology Edin. 1946.)

Poppelreuter's observation — which I believe has not been anatomically verified — of 'amorphous' reactions to light in blindness in the hemi-anoptic field or after trauma or disease of the occipital pole might also provide some support for this view (quoted from Wald and Burian, Am. J. Ophth., 1944, V. 27, 950.), although disturbed associative connections might also be possible in this case.

In conclusion a few remarks on the pupil reaction. With normal children one tends to regard the reaction of the pupil as a sign that the optical tract is intact in its entirety. In our cases of myelination disturbances we have also usually regarded the reaction of the pupil as a sign that the child 'saw'. Magitot even saw some sign of a pupil reaction in premature infants (5 and 6 months) and opposed on these grounds the opinion of Beauvieux that nondetectability of the myelination process in the optic nerve meant that the optical stimulus could not be conducted further. Since these infants had all lived one or more weeks before being examined by Magitot, and since light stimuli are believed to stimulate the myelination strongly, Beauvieux considers that the possibility of the somewhat worm-like contractions of the pupil in these children being due to a more or less normal conduction of the light stimulus is not in conflict with his view. [1])

The myelination of the fissura calcarina is complete at birth. But it is doubtful whether this is already the case in the 5th. month of gestation. The stimulating function of the light stimulus will make itself felt first in the optic nerve, then gradually reaching the occipital cortex. It is conceivable that before the corticopetal tract is completely formed a connection of the retina with the subcortical ganglia (corpus quadrigeminus ant. as homologue of the tectum opticum) may already have been established directly or indirectly.

Experimental data show that in dogs, cats and monkeys the pupil reflex is maintained, in addition to the light-perception, after exclusion of the cerebrum. In view of the facts that this reflex was also frequently present in our children with myelination disturbances and in prematures, at a time when no signs of cortical perception could yet be found, and that in many newborns the reaction of the pupil was very slow and the duration of the contraction remarkably short, even when the light stimulus was continued, it is here also permissible to assume that this first reaction of the pupil was brought about in the same way as in the mammals. In the course of continued growth the process of encephalization will gradually transfer this function also to the cortex.

As the observations in cortical blindness show, however, the pupil reaction is not inseparably linked to the cortex.

On the grounds of anatomical and physiological data and the results of our investigation we come finally to the following conclusions:

    (1) Both the central and the peripheral parts of the organ of vision are still incomplete at the time of birth.

---

[1]) At this very early stage the possibility of a direct action of light on the iris also deserves consideration.

(2) In particular, the myelination of the central optic tract still shows many gaps and imperfections, which result in disturbances and defects in the conduction of stimuli.

(3) In consequence of these disturbances and defects the optomotor reflexes will be absent or imperfect.

(4) Since the establishment and development of the reflexes is partly dependent on the myelination and cytodendrogenesis and vice versa, these 2 processes are in turn stimulated by function, the reflexes — at any rate the higher — cannot be developed to any extent, if at all, at the time of birth.

(5) In connection with the very gradual postnatal development of the area para- and peristriata, the building up of the associative system can only take place very slowly.

(6) In view also of the development of the other parts of the brain, there cannot be any question of conscious seeing for the first few weeks at least. There is probably a reactive response, adapted to prevailing conditions, to light-stimuli which are not yet identifiable, this gradually passing into protopathic sight in the course of the first few months.

The perceptive, the optomotor and associative components of a baby's sight thus leave a great deal — if not everything — to be desired at the time of birth. The question as to the infant's sight asked at the beginning of this chapter can, in our opinion, be best answered by the remark that on the whole we have to take this 'sight' with a grain of salt and have no justification for believing it to be equivalent or even comparable to the sight of the adult.

The correctness of the observations of Donders, Engelmann and Hering is in no way doubted; only the interpretation may change in the light of the time.

CHAPTER VIII

AMBLYOPIA

A chapter on amblyopia might be dispensed with in a study concerned with the aetiology of squint, were it not for the fact that in the history of the problem amblyopia has always occupied — and still continues to occupy — an important position in theories on the aetiology.

The correlation between squint and amblyopia is obvious, but no answer has yet been found to the question of whether amblyopia is the cause of squint or squint the cause of amblyopia, or whether both proceed from a common cause.

On this point Van der Hoeve writes: 'There exists a voluminous literature, the authors of which — although they include some of the most eminent members of our profession — have not always remained objective. Many adopt an attitude of unitarism: They assume the existence of primary squint and reject the possibility of primary amblyopia altogether, or they reject primary squint and show themselves protagonists of primary amblyopia.'

This judgement emphasizes once more the confusion and uncertainty still prevailing in the problem of squint.

A correct definition of the concept amblyopia must precede any discussion. With the improvement of our understanding of its cause and nature both anatomically and physiologically, the definition has been made sharper, but it is certainly still impossible to speak of a generally accepted solution of the problem amblyopia.

As regards the anatomical definition, the results of the investigations of the field of vision showing the presence of a relative or absolute scotoma (Heine, Evans, Peter, Haitz and others) have contributed to the more or less general conviction that the disturbance is confined to the macular field, although it was originally considered that a general diminution of the perception of form and colour over the whole field of vision, except in the extreme temporal crescent, had been observed (Barrie, Feldman and Taylor).

The results of physiological investigation (Harms, Vogt) supported this view. It was also greatly strengthened by the findings of Wald and Burian, recently confirmed by Blet, Rubino & Pereyra and others. The electroretinogram in amblyopia (Karpe) points in the same direction.

The above-mentioned investigations also showed that amblyopia must be regarded physiologically as a disturbance of the perception of form, while the light and colour senses are normal both peripherally and centrally. Is there, however, an anatomical basis for this affection or is it of functional character?

No general agreement on the answer to this question has yet been reached. There still exists a group in which, on the authority of Priestly-Smith, v. Graefe, Poulard, Smukler and others, it is assumed that haemorrhages in the macula at birth may constitute the (afterwards ophthalmoscopically undetectable) cause of amblyopia or, — as suggested by Clark — that an atrophy of certain cells of the corpus geniculatum laterale might develop as a result of the absence of functional stimuli from the retina and the cortex.

Although Uhthoff was able to show that macular lesions occur, permanent damage as a result of this is considered on various grounds to be rare. In contrast to these rare cases of anatomical or passive amblyopia is the concept of an active amblyopia, present in the overwhelming majority of strabismus cases and believed to be of functional origin. Unlike the passive form, this active amblyopia — which generally develops before the 6th. year — is nearly always capable of improvement, as shown by the investigations of Worth, Juler, Meller, Peter, Miranda and Travers and, in particular, by the large-scale research of Bielschowsky and Sattler.

From my own experience I can confirm the results of these investigations unreservedly.

On 984 cases of squint, amblyopia was found 416 times. All these cases had a unilateral squint. Cases of strab. alternans are, thus, not included, nor are those with an objectively detectable macular lesion.

Of these 416 patients 224 were aged 6 yr. or younger: 192 were older than 6 yr. In all patients under 6 yr. the amblyopia was seen to improve, even in cases in which it had probably developed before the 1st. birthday. Amblyopia in infants under 1 yr. is difficult to detect directly; as a rule it can only be deduced from the fact of the exclusive use of one eye. I feel justified in stating it as my conviction that in very joung children there practically never occur cases of true amblyopia which cannot, with careful treatment and exercises, attain at least an average visual acuity. This is quite easy to understand in the light of the nature of amblyopia, to which we shall shortly return.

In view of the great rarity of such cases, it might be as well to drop the distinction between active and passive amblyopia.

If, however, one desires to use the term amblyopia to mean only a functional maculo-cerebral disturbance characterized within certain limits by reversibility, it is no longer justifiable to apply this term to cases in which the disturbance of vision is due to anatomical abnormalities of the macula and its central connections. But in the literature — even in the literature on strabismus — we find it repeatedly used to denote any kind of decrease of visual acuity, irrespective of its origin. It would be highly desirable, in the interests of correct understanding and accurate terminology, to agree to reserve the term amblyopia exclusively for cases of impaired visual acuity in which no anatomical lesion of the maculo-cerebral system is detectable.

Apart from deviations in the position of the eyes, the only factors to be regarded as possible causes would then be congenital or early-established clouding of the media, while under some conditions aniseiconia or certain serious errors of refraction might have to be taken into consideration.

Is it permissible, on the grounds of the foregoing considerations, to describe the 'sight' of the newborn as amblyopic, as is often done (Duke-Elder)?

It is now generally agreed that the macula possesses, thanks to its special structure and its connections at higher levels, certain extra possibilities in addition to the ordinary qualities of sight.

The investigations of Wald and Burian now demonstrated that amblyopia is in fact based upon a temporary or permanent loss of these higher maculo-cerebral functions.

During the first 6 months the infant will develop protopathic sight, which will then gradually acquire the qualities of our peripheral sight. During this time the macula is still passing through its anatomical development and, therefore, it is not yet possible — at any rate in the first few weeks — to speak of higher macular functions. For this reason it would be incorrect to apply the adjective amblyopic to this primitive, dyscritical acuity of vision in which macular epicritical sight does not yet participate. The term amblyopic does not characterize the magnitude of the power of sight but rather indicates that a developing or already established vision has suffered a decrease of chiefly functional origin.

The contrast between the different views is most clearly evident when we come to consider the primary or secondary character of amblyopia, and here also the conflict touches the problem of the origin of spuint.

As we have already seen, the number of cases in which amblyopia is the result of haemorrhages during birth, or possibly of nuclear aplasias, is considered to be extremely small, although there is no doubt that such cases do occur. But according to our definition we are not really justified in calling these cases 'amblyopia'. Not only are anatomical lesions detectable, but in these cases of so-called congenital amblyopia the epicritical sight will never be able to develop — for reasons other than functional. In these cases there will be from the time of birth onwards a central scotoma in one or both eyes, with a high probability of secondary strabismus.

Statistics show, however, that the occurrence in connection with squint of such cases of poor central visual acuity, not amenable to improvement, is very rare.

From the fact that Rubino and Pereyra found a normal light-sense in 20 cases of 'congenital amblyopia' it can be deduced that eyes with 'congenital amblyopia' behave from a physiological point of view in the same way as those considered to be affected by 'secondary amblyopia'. The separation of the two forms thus appears to be unfounded.

Another possibility which might call for consideration is a congenital, hereditary form of amblyopia. A strong argument against a genetic origin is, however, the very common unilateral occurrence (Heinonen and others). Hereditary anomalies of a bilaterally-symmetrical organ are usually, or at any rate frequently bilateral. But bilateral amblyopia is not found in squinting families. Waardenburg comes to the conclusion that 'pathological studies on twins have proved beyond doubt that amblyopia is not the cause of squint, neither is it the obligatory consequence thereof'. In connection with the last statement, the dual nature of the concept amblyopia should be kept in mind.

Finally, the simple dominant hereditary transmission of squint also pleads against amblyopia as its cause.

The opinion that the amblyopia is only secondary, and may thus be a consequence but never a cause of squint, is confirmed in various ways.

As we have already pointed out, the term amblyopia cannot — for anatomical reasons — be applied until the age of 6 months has been passed.

The following argument, derived from practical experience, is based on the large number of spontaneous cures of strabismus, which careful observation reveals in cases of amblyopia even after the 6th. year of age.

A rather high percentage of the ordinary refraction cases shows a decreased visual acuity in one eye for which no cause can be found. The ophthalmologist who examines candidates for military service finds that a considerable proportion of the cases of uni-ocular reduction of visual acuity have to be placed in the amblyopia group. At first sight there is nothing in these cases to suggest a previous squint and in some it is not even possible to find any trace of heterophoria. A careful anamnesis, which should include also the near relatives, will serve, in view of the high degree of hereditary occurrence of squint, to put us on the right track. Where the affection has been unnoticed by the patient or his relatives, the existence of familial cases affords a high degree of probability to the diagnosis. Childhood photos often abolish doubt in such cases. I have repeatedly been able, on the grounds of these objective data, to convince a military medical board that the possibility of simulation could be excluded.

In many cases, however, it will be necessary to proceed along different lines and to make use of other methods in order to establish the diagnosis of a previous squint. Tests for:

(a)   a normal central colour sense;

(b)   the presence or absence of a normal adduction reaction of the amblyopic eye on fixation of an object at a short distance with the eyes uncovered;

(c)   the existence of a dissociation in following movements in the horizontal field of gaze with both eyes at once, repeated in the dark-room if necessary, will help, in addition to the Rodman Irvin test for binocular vision in doubtful cases, to clinch the diagnosis.

Upon examination along these lines it will be found that a

surprising number of amblyopic patients without squint have indeed squinted as children, so that a spontaneous cure of squint must be a much more frequent occurrence than the textbooks would lead us to believe, while this cure must have taken place during the existence of the amblyopia.

Thus, squint has a high tendency to spontaneous cure, not only in early youth, as we have already seen in the frequency curve in Chapter II, but also in later years.

If we now examine our own cases, with regard to the connection between amblyopia and squint, taking at random 190 patients under 10 yr. of age in whom amblyopia associated with unilateral squint was observed, we find that the parallel position of the eyes had been attained at a given time in 90 (rather more than 71%) of the 126 cases diagnosed before the 6th. year, while at that same moment the amblyopia had disappeared in 70 children (rather more than 55%).

Of the remaining 64 patients, between 6 and 10 yr. of age, 42 (nearly 66%) had at that same moment a parallel position of the eyes, while only 11 (rather more than 17%) had lost their amblyopia. (See table and graph below).

Relation between AMPLYOPIA and position of the eyes
(Summary of 190 cases with strabismus convergens unilateralis)

Year in which the amblyopia was diagnosed	No. of cases	Amblyopia cured	Amblyopia not cured	Eyes in parallel position
1st.	2	1	1	2
2nd.	5	3	2	3
3rd.	26	19	7	20
4th.	31	18	13	22
5th.	38	19	19	28
6th.	24	10	14	15
7th.	26	6	20	18
8th.	18	5	13	13
9th.	9	0	9	5
10th.	11	0	11	6
Total:	190	81	109	132

A graph can elucidate this relation.

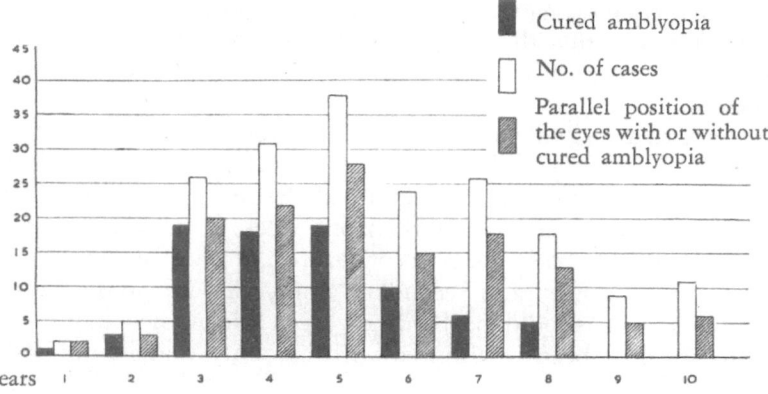

Relation between amblyopia and position of the eyes in 190 cases
with strabismus convergens unilateralis (see table).

This brings to light the remarkable fact that the parallel
position of the eyes was attained in about 2/3 of cases in each
age-group, irrespective of the number of cured amblyopias, the
latter actually being more than 2½ times as large in the first
as in the second group. Further, we see that the lack of a good
central visual acuity in one eye, and hence the possible absence
of central fusion, does not necessarily constitute a hindrance to
the attainment and maintenance of a parallel position of the
eyes. It would appear, thus, that other factors play the decisive
rôle here.

It can be further deduced that the treatment of amblyopia,
however useful and necessary this may be for other reasons,
does not appear to have a decisive influence on the eye position
finally attained. Both these conclusions are very important also
from a therapeutic point of view and seem to point to a fundamental error in the methods so far employed in orthoptic
training.

We also note that amblyopia could be detected in only 2
cases under 1 yr. of age, that this number was still as low
as 5 in the 2nd. year and that in the 3rd. year it rose suddenly to 26. After this the number still continued to rise, but
in view of the fact that all kinds of factors (treatment, social
circumstances etc.) now come into play and influence the
figures in a way which is difficult to judge, I prefer to leave
the subsequent figures out of consideration.

The frequency curve of squint showed a peak during the first year. The fact that the peak for amblyopia does not appear until 4 years later pleads unequivocally for the secondary character thereof.

As a final argument we may recall another well-known fact: that in very many children under 4 yr. of age it is possible to produce an amblyopia (alternating if desired), by covering one eye, sometimes even in less than one week. This would appear to complete the case for a secondary occurrence of amblyopia. The terms primary and secondary amblyopia are productive of confusion and should be banished from the nomenclature.

It has now been established with certainty that in the overwhelming majority of cases amblyopia can be abolished by timely and correct treatment.

As can be deduced from our survey, and as confirmed in practise, the chance of recovery — which so far was still relatively great — drops rather suddenly after the 6th. year and becomes very small after the 8th. year. These values, however, must not be regarded as absolute, since they are influenced — usually in an unfavourable sense — by various other factors which have nothing to do with the recovery process itself. School-attendance, for instance, makes the rational prosecution of training practically impossible, while the chance of recovery is in direct proportion to the perseverance and care with which the parents and the doctor carry out the training. Our figures, therefore, represent only the results attained and not those attainable.

For rational treatment it is necessary to cover the non-amblyopic eye with a dressing fixed with sticking-plaster and to keep this in place day and night during the first stage of treatment. The duration of treatment must be decided individually and depends on the child's age and on the existing conditions. As the visual acuity increases, the bandage can be left off intermittently, at shorter or longer intervals, which again must be determined individually. It must be borne in mind here that the bandage blocks not only light but also the passage of possible motor impulses, evoked by the light stimulus, from the good eye to the amblyopic one. Furthermore, not only the central but also the from peripheral parts of the retina evoked optomotor reflexes of the amblyopic eye are exercised, in particular also the abduction reflexes.

In view of the maturation process of the brain, the training should theoretically be continued up to the age of about 12 yr. In practice, however, this cannot be done and the training is generally suspended at the age of about 9. The earlier the treatment starts the better and more certain is the result.

Of the 126 children under 6 yr. in our survey, 55% had attained an improvement in their amblyopia at a given time. For the 64 over 6 yr. this proportion was only 17%. The later one starts treatment after the 6th. year, the more difficult it is to get a result, even with prolonged training, while the visual acuity finally reached is generally much poorer than that achieved with younger children.

In some of these children I noticed that the visual acuity for distance was relatively low in comparison with that for near objects.

In the literature it is usually stated that exercise of the amblyopic eye can never lead to a visual acuity higher than that which the eye possessed at the time when it became amblyopic. (Worth, Chavasse, Keith Lyle). This would thus give a very bad prognosis for children whose amblyopia had started at a very early age: their amblyopia would in fact be incurable. On the grounds of personal experience, however, I am compelled to oppose this view. I can without any difficulty pick out about 20 children who had become amblyopic before their 3rd. birthday and who after treatment acquired a visual acuity of at least 4/5.

I am also unable to share the current opinion with respect to the maximum generally attainable visual acuity in children of a given age. Chavasse gives a table for this.

Every child whose treatment is started in good time can attain a visual acuity equal or approximately equal to that of the good eye, provided that the conditions are the same on both sides. The attainable visual acuity depends not on the age at which the amblyopia appeared but on the age at which treatment is started. The visual acuity attainable at a given age varies individually. A value of 5/5 is no exception at the age of 4 yr.

As we have already remarked en passant, the time necessary for the achievement of this result is greatly dependent on the age. With very young patients (2-3 yr.) much can be achieved in 2-4 weeks; with children aged 3-5 yr. the same number of months is often needed. The required duration of treatment and the result attained are further dependent to a high degree on the exactitude with which the treatment is carried out. For this purpose it is essential to secure the co-operation of the parents. In the course of this collaboration the oculist must demonstrate to them from time to time the improvements in the originally poor visual acuity, and the child must be rewarded. Such a procedure is stimulating for all parties. With children under 6 yr. the attainment of a visual acuity of 4/5 is the general rule, and this should be made the minimum target.

It has been repeatedly stated — and also observed by the writer himself — that amblyopia may improve at a later period of life (Van der Hoeve, Jaensch, Hartmann and others, as quoted by v. d. Hoeve in the Symposium on Strabismus, 1943). Nearly all were cases in which the sight of the good eye was lost and stern necessity compelled the patient to exercise his amblyopic eye. One is forced to assume that in these older adults the poss-ibility of improvement was latent owing to the fact that their amblyopia had first developed at a relatively late age, or that their squint had for a long time been only periodic.

In both these types of case there has existed for some time a possibility of development of the macular functions during the period in which that normally occurs. It is generally believed that this period ends about the 12th. year. Nevertheless, we prefer to leave open the possibility that even in later life an improvement of the original visual acuity can still occur, by means of certain changes in the synapses and the formation of new associations.

Chavasse's rule that the attainable visual acuity is not higher than that existing at the time when the amblyopia began is, therefore, probable also inapplicable to these older, exceptional cases.

One of my own patients, a teacher who underwent at the age of nearly 80 an operation for cataract on her one remaining, amblyopic eye, reached after the operation a visual acuity which permitted her to do embroidery.

Once a good visual acuity has been reached by training of the amblyopic eye it is generally maintained. Regular re-examination is, however, desirable in certain cases.

The longer the time that has been required for the training the less will the eye be liable to fall back into its old bad habits when left to itself. The child's age is here again an important factor. When the good eye is covered before the 4th. year, the possibility of amblyopia developing in this eye must always be reckoned with. The younger the child the sooner this occurs.

It can however be abolished with equal rapidity by alternating the covering.

Repeated re-examination is especially necessary in these alternating cases.

In connection with the character of amblyopia as outlined above, we may assume that it will generally be present in cases of strabismus unilateralis. For cases which have developed before the age of 2 years this rule seems to hold without exception. If a case of strab. alternans passes into one of unilateral squint, it may occur, especially with rather older children, that the squinting eye will continue for some time to show a visual acuity which is unusually high for amblyopia. The same applies to cases of strab. periodocus becoming continuous at a later age.

Some attention to the problem of the cause and mode of origin of amblyopia and the central field to which this seems to belong is certainly not out of place here.

The literature and the text-books and handbooks say remarkably little about it; most of them merely speak of amblyopia 'ex anopsia' or as a result of suppression and do not go further into the mechanism.

If we regard amblyopia as secondary and thus as a consequence of preformed agencies, it seems the obvious thing to ascribe it to the disturbance of optomotor reactions — and hence of binocular single vision — which occurs in squint. The conflicting or incompatible impulses produced in this way will, as Zeeman has pointed out, be annulled by inhibitions. Such inhibitions are constantly occurring in daily life. Various optomotor reactions, both monocular and binocular, will be restricted by them in their development or may even be altogether prevented from developing. The same applies to the cerebral processes correlated with these motor reactions. Disturbed emmetropisation must also be included in this group.

We can imagine sight as being built up of various components: a perceptive, an optomotor and an associative. An inhibition of motor impulses will inevitably have a braking effect in the perceptive and associative sphere, with disturbances in the function and also in the morphological development as results.

With a given degree of development of perception and association the visual acuity will be determined in the first place by the degree of sharpness and fine gradation of the optomotor reflexes.

I gladly endorse Roelofs' opinion that the best visual acuity will be present when each individual retinal element is connected with a sharply regulated optomotor stimulus. This regulation develops during function. The high visual acuity of the macula can be achieved only by exercise of the relevant optomotor reflexes (the fixation and fusion reflexes). In correlation with this proceed the building up and maturation in the perceptive and associative spheres and also the development of the refraction (emmetropization). If the development of these reactions is disturbed or prevented by continuous or constantly repeated inhibition, the finer reflex regulation will not be able to develop. If the connections for properly regulated optomotor stimuli have only recently been established, they may be lost again as a result of covering or inhibition of the eye in question. Inhibition, thus, disturbs the chances of a normal development of the visual acuity and in this way constitutes the cause of amblyopia.

As we have already remarked, amblyopia is probably related to the cortex. The evidence for this central situation has been provided chiefly by the work of Wald and Burian. If we seek the cause in an inhibition of the optomotor stimuli and if we may assume that the large pyramidal cells in the limes parastriata play a part in the production of these stimuli (Economo), it is conceivable that the 'suppression area' that McCulloch believes to be situated at the outer margin of the area peristriata might have a rôle in the establishment of these inhibitions. All this, however, is still very uncertain.

From our considerations on the nature of amblyopia it has already emerged that both the development of function and that of the morphological structure are disturbed as a result of the inhibition.

The disturbance in amblyopia is, thus, both functional and structural and amblyopia is a dual concept.

It will depend upon circumstances whether it will take the form of a continuous or an intermittent disturbance. Clinically it is certainly necessary to distinguish between these 2 forms of amblyopia, as is in fact done in the larger handbooks. If the inhibition of given optomotor stimuli is a continuous process there will be, in addition to a disturbance of function, an imperfect morphological development as result. Therefore this form of amblyopia will also persist without inhibition. The literature of the Englishspeaking countries calls this 'amblyopia of arrest'.

If the inhibition is intermittently present, e.g. in strab. alternans, the structural development may be normal. leaving only a functional disturbance. Here the term 'amblyopia of extinction' is used. The terms 'structural' and 'functional' might also be used.

We can represent the cause and nature of amblyopia schematically as fellows:

Inhibition ——————— inhibition of optomotor stimuli ——

insufficient exercise ⟨ disturbed function

deficient morphological development.

As an optomotor disturbance, amblyopia manifests itself as a disturbed seeing of form. It goes without saying that the light and colour senses may still be entirely intact.

These views of Roelofs are entirely in agreement with the results of studies of the light and colour senses of the amblyopic eye by Vogt (1939), Wald and Burian (1944), Rubino and Pereyra (1948) and Blet (1950).

The fact that all authors are unanimous in describing the light perception and localization as normal, while Wald and Burian also find the colour sense to be entirely intact, both centrally and peripherally, pleads strongly — as does also the normal electroretinogram — against the ideas of Harms and others as to the existence of a retinal inhibition. This view is also opposed, as we have already pointed out, by Wald and Burian. In the recently-published, revised edition of Worth and Chavasse's 'Squint', however, Keith Lyle still retains it. Roelofs' views are in perfect accordance with the cortical theory.

We may now ask whether amblyopia is a significant factor from a social point of view and whether it may perhaps be described as a social evil. A statistical survey recently published by Downing provides the answer:

On a total of 60,000 recruits he observed 1,920 cases of amblyopia, including:

855 cases of uniocular amblyopia without strabismus (1 : 70)
770 „   „   „   „   with. strab. convergens (1 : 78)
295 „   „   „   „   „   „   divergens (1 : 208)

If we neglect the — undoubtedly considerable — number of cases of cured squint included among the 855 of the first group, Downing's figures mean that 1,065 cases of strabismus with amblyopia were found among 60,000 men called up for military service, i.e. rather more than 1 : 60 (1,8 %).

In view of the fact that this was a selected group of a certain age, from which certain categories (e.g. mentally defective and diseased individuals) were automatically excluded, and which consisted entirely of males, we may conclude that an estimated incidence of 2.2 % over the whole population will not be far from the truth.

If we also consider Downing's conclusion that 66% of all cases of uniocular loss of sight are due to amblyopia, we may justifiably be surprised that such a serious handicap to the individual and such a loss to the community of working ability has been tolerated without protest for so long. There seems to be an important task awaiting the attention of preventive medicine and of The International Association for the Prevention of Blindness.

The conclusions pronounced by Van der Hoeve in the Strabismus Symposium in 1943 on the subject of amblyopia ex anopsia, and the considerations advanced by him in this connection, have lost nothing of their validity to day. His remark: 'The high incidence of strabismal amblyopia constitutes and enormous loss of the power of sight', is once more clearly confirmed by Downing's statistics.

Summing up we come to the following conclusions:
(1) Cases of primary (congenital) amblyopia are statistically rare. The eye with 'congenital amblyopia' behaves physiologically in the same way as that with 'secondary amblyopia'.
(2) The amblyopia associated with squint has a secondary character and is up to a certain age always amenable to improvement.
(3) Amblyopia is the result of imperfectly regulated optomotor stimuli and is probably related to the cortex.
(4) Amblyopia is as such not hereditary.
     It may be not only functional but also structural.
     The treatment of amblyopia is of social importance.

Both squint and amblyopia are due to a disturbance in the development of the optomotor reflexes: the from peripheral parts of the retina evoked reflexes in the first and the central reflexes in the second case. The cause, however, is different in each case.

On the grounds of these conclusions the answer to the question asked at the beginning of this discussion can only run as follows: amblyopia is not the cause of squint but squint must be regarded as the cause of amblyopia.

CHAPTER IX

NEW VIEWPOINTS AND HYPOTHESES ON THE ORIGIN OF SQUINT

Reflexes are the elemintal units in the
mechanism of perpetual equilibration.
Pavlov.

I PHYSIOLOGICAL SECTION

Object and significance of reflexes.

NOTHING comes from nothing. In a material sense there is no existence and no rest unless there is equilibrium between forces acting internally and externally.

In the realm of life this material equilibrium is not sufficient. For its preservation life demands the maintenance not only of a material equilibrium but also of a biological equilibrium, based on harmonious interaction with influences from the environment.

Stimuli from the environment are continually reaching the individual; these disturb the harmony and lead in a reflex manner to a reactive event tending to restore the disturbed equilibrium.

As with vital phenomena in general, this effect is directed towards the preservation of the individual, so that a given stimulus must always be followed by a given reaction, in accordance with a definite law.

More than 3 centuries ago Descartes grasped this connection between stimulus and effect and realized the great importance of such reactions. His idea that every reaction of the animal organism was the inevitable consequence of a preceding stimulus; that the relation between stimulus and effect was established over certain pathways and that the nervous system to which these pathways belonged existed for the sole purpose of establishing such connections, could still be used nowadays as a definition of the term 'reflex'.

Sherrington, Magnus and Pavlov also took these fundamental

views as the starting point for their investigations. Pavlov further classed the reflexes as elementary unconditioned reflexes present at birth (inborn reflexes) and a group, acquired at a later date, of conditioned (acquired) reflexes, the latter being always grafted onto the former.

In the ascent through the animal kingdom the number of stimuli received — especially via the telereceptors (nose, eye, ear) — steadily increases, while the increasingly complex conditions of life render the conditioned reflexes (signalizing stimuli!) more and more important. The anatomical substratum has to adapt itself, if the dynamic equilibrium between the individual and the environment is to be maintained. In a physiological sense also, the possibility of adaptation to altered conditions must always remain open, so that it is soon no longer enough that a given stimulus should be followed only by one definite reaction.

Under altered conditions it is necessary that the same signalizing stimulus should now be able to arouse different reactions, while, conversely, the same reaction must be capable of being aroused by different stimuli. Pavlov considers the nervous substratum for this to be provided by the extremely complex structure of the hemispheres.

'So infinitely complex, so continuously in flux, are the conditions in the world around', he writes, 'that that complex animal system which is itself in living flux, and that system only, has a chance to establish dynamic equilibrium with the environment. Thus we see that the fundamental and the most general function of the hemispheres is that of reacting to signals presented by innumerable stimuli of interchangeable signification'.

Experiments show that loss of the hemispheres renders the animal unfit for an independant existence and that the greatest care is needed if such an animal is to be kept alive.

This objectively demonstrates the very great importance of reflexes for the preservation of life.

If we agree with Zeeman's view that the boundaries between unconditioned, conditioned and voluntary reflexes have been shown by Pavlov's research to be highly artificial, we may almost be inclined to ascribe a reflex character to all vital manifestations or to regard these — in so far as they are mental expressions of life — as a phenomenon accompanying the reflex or as the psychological correlate of the impulse aroused. (e.g.: localization).

As pointed out in Chapter I, the older physiologists were convinced of the reflex character of the reactions observed in the eye. They considered that the idea of voluntary events could be practically excluded in this connection. These views have been confirmed by modern research and the number of scientists now accepting them is daily increasing.

I do not believe that anybody nowadays doubts the completely reflex nature of the ocular movements of babies and young children. In accordance with the state of development of their hemispheres, the number of signalizing and other conditioned reflexes — which, moreover, are to some extent still in the stage of being built up — is small, so that the elementary reactions are more clearly in evidence. Very young children thus, are ideal subjects for the study of such reactions.

A consequence of the views set out above is that a reflex is always the obligate consequence of the disturbance of (static or dynamic) equilibrium, but it follows equally that every reflex must be preceded by the disturbance of this equilibrium by a stimulus.

Every state of rest, thus, is the expression of a biological equilibrium; a change in this state can occur only under the influence of a precedent stimulus.

Nothing comes from nothing.

Ocular postural reflexes: If we attempt, in the light of the foregoing considerations, to ascertain which stimuli will — after birth — have a decisive influence on the biological equilibrium position of the eye, we find in the first place that in this period — in contradistinction to that which preceded it — it is in fact permissible to speak of a position of rest. As we have already seen, the eyes immediately after birth are still in almost ceaseless movement, but once they have been spontaneously opened and the light has aroused the slumbering potentialities of the retina they acquire direction (a definite position) as a result of this adequate stimulus. Then, seeking equilibrium between the internal and the new, external forces acting on them, each eye separately gradually assumes its biological position of rest.

This interplay of vital forces can be actually observed in the newborn. If the eyes open spontaneously they seek their equilibrium position. If the light stimulus is excluded, each eye independently resumes its slow wandering movements.

The possibility that in certain cases abnormal anatomical

conditions may form an insuperable obstacle to the attainment of the normal position of rest has to be mentioned here; such cases, however, are exceptional. The attention paid to them was a consequence of the mathematical-mechanical view which for so long has dominated the question of squint, but which should now be replaced by a biological view. It goes without saying that in a later period of life the position established may be influenced, in a positive (corrective) or negative sense, by other biological factors. We shall deal with these later.

If we now attempt to define the chief stimuli which determine the position of the eye under various conditions, we come — in ontogenetic succession — to the following arrangement:

(1) idiomuscular (biophysical) stimuli;
(2) proprioceptive stimuli (possibly also enteroceptive);
(3) vestibular stimuli;
(4) musculo-sensory stimuli;
(5) sympathetic stimuli;
(6) sensory stimuli, including
(7) light stimuli.

Stimuli 1, 2, 3, 4 and 5 give rise to reflex eye movements even before birth. The idiomuscular stimuli, arising in the ocular muscles, must be conducted by the muscle fibres themselves. They give rise to slow, invariably monocular movements. It appears not unlikely that they play some part in the slow wandering movements. They may occur in company with nervous stimuli. The elasticity of the skin, muscle and tissues is perhaps the earliest factor responsible for movements (Minkowski).

Proprioceptive reflexes are evoked by stimuli coming from the proprioceptive sphere; (here we are concerned particularly with the proprioceptors of the ocular muscles and the surrounding fasciae.) They are always monocular. The monocular proprioceptive reflexes remain highly important throughout life and are indispensable for the creation and maintenance of binocular vision.

From a quite early time we find in the literature (Petrus Camper, Worth) the idea of a connection between enteroceptive stimuli and squint. This is, however, by no means certain.

The vestibular reflexes, which are always conjugated, are certainly present by the end of the 5th. foetal month in man (Bartels and Zika), while the vestibular apparatus itself is already completely differentiated in a 4 cm. embryo (Minkowski). They continue, based upon the general postural reflexes, to play their important part throughout the individual's life. While in the higher mammals they are still of predominant importance in the compensation of the position of the eyes upon movements of the head (Van der Hoeve, de Kleyn), in apes and man they have lost much of their importance in this respect, on account of the new demands made by binocular single vision.

They are, however, of the greatest importance as a basis for the construction and development of the binocular optical reflexes.

The large group of musculo-sensory reflexes includes the cervical reflexes; these are important for our purpose. They are always conjugated and are closely related to the vestibular reflexes, probably having a large part of their path in common with these. In the newborn they are still clearly detectable (Bárány), but at later stages of life they reappear only under pathological conditions (Stenvers).

The sympathetic and sensory (acoustic) reflexes are of little interest for our purpose and I do not propose to discuss them.

All these forces act on an organ which has had its anatomical position of rest mapped out for it, within certain limits, by its bilaterally symmetrical origin and development.

Physiology of the ocular movements (subcortical reactions).

As a result of the stimuli listed above as 1 to 5 inclusive, the eyes of the foetus before birth will be able to perform 2 kinds of movements: monocular and conjugate movements.

If we imagine the head and trunk to be fixed, we can expect only monocular movements, as both the vestibular and the cervical reflexes will then be out of action, at any rate in so far as the foetus itself is left at rest. With a conjugate movement both eyes receive simultaneously a stimulus to movement in the same direction, with a synchronous bilateral movement as result. We must beware of thinking of this as a binocular movement based upon a definitive and ideal junction between the 2 eyes. Both eyes are merely obeying a given impulse to movement, independently of the position which they are at that moment occupying with respect to each other, the amplitude of the movement being determined by a movement of the head (vestibular reflex) or a movement of the trunk with the head fixed (cervical reflex) or a combination of the two. There is still no question of linking between the 2 eyes in a functional sense, nor does it appear that such linking would be of any service to the individual at this prenatal stage. This is not included in the phylogenesis and can derive its raison d'être only from optical demands in the postnatal period.

As we saw in Chapter VII, these theoretical considerations found confirmation in the examination of the eye movements immediately upon and shortly after birth; (observations in the Groningen Obstetrical Clinic). Immediately after birth of the head the eyes were covered with a damp bandage and the infant was taken, the moment it was completely expelled, to the dark-room. There, with the weakest possible red light, a fragment of the membrane of a raw hen's egg, to which a little luminous paint had been applied, was placed on each cornea. This adhered well but could afterwards be easily dislodged by pushing with the eyelid. The eye movements were now studied with ultra-violet light and it was seen that the fluorescent points of light moved like an 'ignus fatus', very slowly and apparently aimlessly to and fro in a completely dissociated manner, so that the patient observer was rewarded with every possible position of the eyes, from an excess of convergence to the most extreme bilaterally divergent position. Turning of the head to the right or left was followed by a prompt compensation reaction. With move-ment of the head in a sagittal direction this was less pronounced. The influence of cervical reflexes was not investigated.

The light stimulus being excluded, the dissociated, monocular, very slow wandering movements, observed also with head and trunk fixed, in the dorsal recumbent position and at rest, can be regarded as reflexes produced by stimuli which are chiefly idiomuscular or proprioceptive.

If it be the case that the very slow speed of the movements is a sign of a low rate of impulse conduction (this is about 0.1 cm/sec. in non-nervous as against 45 cm./sec. in the most primitive nervous tissue) the notion of reactions to idiomuscular stimuli seems a possible one.

A striking feature in these newborns was, thus, the predomin-antly monocular character of the eye movements, pointing to a total dissociation of the eyes at this period of life.

Not only in newborns was this dissociation of the eyes in evidence, but also in children with serious disturbances in the myelination of the optical system. Their eye movements in the first stage were exactly like those of the newborns (see Chapter VI: Myelogenesis retardata).

In explanation of the type of these movements, attention was there drawn to the analogy between the evolutionary development of general motoricity through the animal kingdom and the various phases in the development of movement in man, as emerging from Minkowski's comparative embryological investigations.

It seems obvious that we must consider these apparently abnormal and aimless eye movements in the light of evolutionary development.

If any further proof were wanted that we are not concerned here with organ-specific movements under the influence of light stimuli, this is provided by Bartels' observation of wandering movements in an infant with anophthalmus.

As a result of the investigations described here and of such further statements as are to be found in the literature, I feel justified in concluding with certainty that the human being begins his existence with completely dissociated eye movements, possible as a legacy explainable on phylogenetic grounds.[1]) At this stage there is, of course, no question of anything like optical linking.

The significance of this dissociated condition for a correct understanding of the physiology of ocular movements, and hence also of the syndrome of strabismus, can hardly be exaggerated. An insufficient realization of the great difference that must exist in all respects between the seeing and looking of the newborn or young baby and that of the adult has had the result that we have delayed too long in drawing a dividing line between observations made on adults and the 'sight' and ocular-movementpattern of the newborn. In this way we have failed to appreciate the prototype which the latter present. Too little attention has been paid to the gradual building up of the necessary reflexes. With and as a result of the semi-decussation of the optic nerve fibres in the chiasm the human being receives no more than the anatomical possibility for a far-reaching association of his two eyes. Not until after the subtle but admirable reflex mechanism has become established, and after practice and experience, will it be possible to create the necessary synaptic junctions which, supported by the anatomical substratum provided for them, will be able to transform the original

---

[1]) Cf. Rochon-Duvigneaud. Bull. et Mém. Soc. franç d'Opht. 1, 1933.

dissociation into an effective association.

Effects of the light stimulus (cortical reactions). All this, however, can be brought into being only in the postnatal period, under the influence of light stimuli. Zeeman writes: 'Up to this time all the eye-movements of the newborn have been able to achieve ,Schaltungen' (synaptic junctions) only by way of the proprioceptors and the sensory nucleus of the trigeminal nerve, thus carving out channels in the oldest cell-complexes of the central nervous system, as far as the thalamus; now from the moment of birth and opening of the eyes, of awakening and exposure to light, they will be accompanied also by the passage of images over the retina and by stimulation phenomena, directed in a particular way, in central retinal representatives. These seem to be predestined to calibrate representations in question and to correlate them with the above-mentioned phenomena in the brain-stem and thalamus, with those anatomical regions which may perhaps be regarded as the entero- and proprioceptive representations of each eye'.

All other bodily movements will presumably help to multiply and strenghten these 'Schaltungen'.

In this way the functions of arousing, activating and warning which belong to the retinal stimulus have added to them the potentiality of following- and fixatingmovements, compensatory head and trunk movements and fixation. The reflexes will become sharper and more exactly regulated the finer the state of development of the sensitivity and the reaction and recovery rates of the light-sensitive elements.'

The effect of the action of light, as a conditioning stimulus will be, thus, that in addition to the reflexes already present, many new ones will develop: the optomotor reflexes, all grafted onto the unconditioned reflexes described in the foregoing.

This development will proceed according to a scheme laid down by ontogeny and will be correlated to the differentiation of retina and macula and the structural building up of the central nervous system in general and of the optical system in particular.

As we explained in Chapters VI and VII, the stage of development which has been reached by the conducting nerve elements, fibrils and myelin sheaths [1]) is of primary importance in connection with the conduction — and in particular with the speed of conduction — of impulses.

[1]) See Le Roy Conel III p. 146.

The individual differences with respect to the stage of mye-lination at birth will be expressed, or revealed, by the optical and optomotor reactions evoked by the light stimulus, with the consequence that these reactions in the newborn may show a wide range of individual variations. In cases of very pronounced retardations the child will be born blind and the eyes will execute certain movements which are proper to their stage of development and are not produced by optical stimuli, as in the 12 cases described in an earlier chapter. (VI)

In less severe cases, the absence or — alternatively — the predominance of certain optomotor reflexes may lead to depart-ures from the biological equilibrium position in one or both eyes, while children without retardation, whose myelination process has proceeded in correlation with a normal general development, will — if born at term — attain a biological equilibrium in the position of the eyes, in which the visual axes are approximately parallel.

In accordance with the dissociation present, however, each eye will seek its own biological position of rest. If the two functionally separated retinal halves have equal motor potential-ities, the forces thence evoked from each pair of diagonally opposite quadrants will approximately balance. The eye will assume a position in accordance with the resultant of these forces brought into action by the motor component of the retinal stimulus. Since we are here concerned with a biological and not a mathematical process, the position thus reached will practically never give exactly parallel visual axes. A very slight deviation (heterophoria) may thus always be expected even in normal cases and in this degree is of no significance for the further development.

Origin of the optomotor reactions. How can we picture to ourselves the coming into existence of the optomotor reflexes?

The 2 kinds of ocular movements, monocular and conjugate, evoked by subcortical stimuli, which we have been able to observe immediately upon birth, may now, after the eyes have opened, also be evoked by light as a conditional stimulus. In this way new possibilities will arise and the number of synaptic junctions may be multiplied, in the manner described above, giving the possibility of adaptation to the conditions and the relations to the external world.

According to Pavlov the conditioned reflexes graft themselves onto the unconditioned (inborn reflexes); Zeeman has extended this idea by assuming that *all* reflexes were once conditional and that they have thus, starting from a very primitive one, developed gradually by a process of grafting of one upon another. In the case of the ocular movements it seems obvious to take as this most primitive reflex that, produced by bio-physical factors in the muscles, (stretching under the influence of gravity when the position of the body is changed) so that in this way the proprioceptive reflex and the resulting monocular eye movement would be the oldest.

The conjugate movements in response to optical stimuli can be conceived of as originating from a repeated concomitance of the passage of images over the retina with movements of the head in space. These movements give rise to labyrinthine stimuli which lead to a conjugate movement in the opposite directions, i.e. in the direction of the image displacements in the field of vision and in the opposite direction to that in which the images pass over the retina. In the course of time these image displacements may become reflexogenic and give rise to a conjugate ocular movement even without a primary head movement or labyrinthine stimulus.

In a similar way, repeated coincidence with musculo-sensory stimuli (cervical reflexes) may also produce such an effect. Thus the optomotor stimuli are not in themselves endowed with motor potentialities but acquire these only conditionally, not-withstanding the degree to which the realization of this potentiality may be facilitated by the anatomical preparation.

Since the extent of movement of the head determines the amount of displacement of the images over the retina, on account of the compensatory eye movements by which it is followed, a calibration will take place with the result that in the course of time this image-displacement will call into existence a proport-ional, oppositely-directed movement of the eyes. It is in this way that it is possible to conceive the origin of the first focussing movement and the later fixationmovements.

How these labyrinthine reflexes may originate and whether these again are perhaps grafted onto the proprioceptive reflexes in the course of their development, I do not propose to discuss here.

The following schemas, worked out by Roelofs, of these monocular and conjugate optomotor reactions, may help to make these clear.

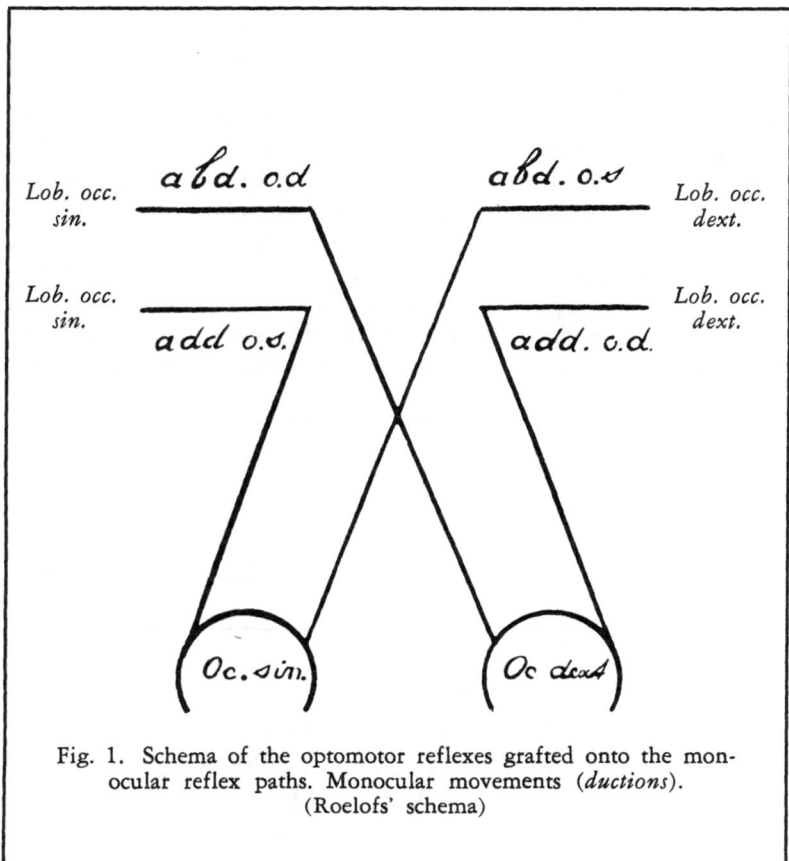

Fig. 1. Schema of the optomotor reflexes grafted onto the monocular reflex paths. Monocular movements (*ductions*). (Roelofs' schema)

Optomotor reflexes in babies. Returning to our babies, we have seen, in the preceding chapter, that the conjugate movements in response to optical stimuli, as described above, are in fact seen on clinical examination of infants in their first few weeks of life. In addition to these, monocular wandering movements — as reactions to proprioceptive stimuli — still regularly occur at this time, although in individually varying degrees.

In this way every possible mutual position of the eyes can be adopted and remain for a shorter or longer time.

The deceptive imitation of squint thus produced by these dissociated positions of the eyes has given rise to the widespread misconception that a true squint is here present and to the conclusion that 'all babies squint'. This 'squint' is, of course, not continuous and is perpetually changing, sometimes from a convergent to a divergent 'squint'. This in itself is enough to convince the patient investigator that the condition is not one of genuine squint, while a rather more prolonged observation would soon have brought to light the highly variable character of these apparent disturbances of co-ordination.

These babies, thus, do not really squint. Their dissociation, however, does give them the possibility of a certain tendency to squint. If in this period their eyes take up a dissociated position, this occurs under the influence of the old proprioceptive stimuli, which formerly evoked subcortical reflexes; in this transition

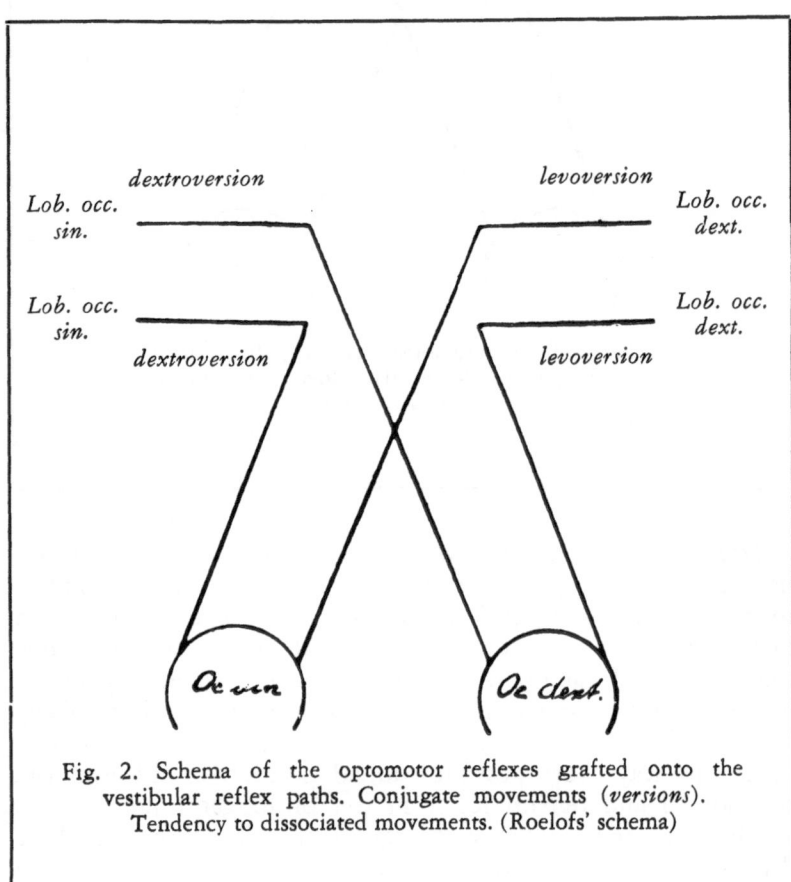

Fig. 2. Schema of the optomotor reflexes grafted onto the vestibular reflex paths. Conjugate movements (*versions*). Tendency to dissociated movements. (Roelofs' schema)

period these can still from time to time predominate over the optical reflexes.

In addition to this, cases may occur in which, during the development of cortical (optical) monocular reflexes, a temporarily excess of predominance of the adduction reflexes over the conjugate reactions may lead to a strong tendency to a genuine convergent squint — i.e. one occurring under the influence of optical stimuli. This will generally be between the 4th. and 6th. month.

The monocular adduction movements which occur from time to time at this stage are the expression of this. They also indicate that dissociation is still present and confirm the impression, derived from the frequency curve for strabismus that all children are predisposed from birth to squint.

The dissociation in question has its consequences. It follows from it that at first each eye will build up its function and optomotor reactions independently, being practically undisturbed by stimuli from the other eye. It will find its biological position of rest independently and will be able to respond chiefly with a monocular reaction to monocular stimuli. From the point of view of innervation, therefore, it has to be regarded as a unit. This agrees with the conceptions of Sherrington. Sherrington's law of reciprocal innervation will be applicable in its entirety to this eye, but Hering's law, which is concerned with a harmonious bilateral innervation, will not be valid for it at this period.

This law is obviously based upon — and perhaps derived deductively from — the association of the 2 eyes, with parallel or almost parallel position, which exists in normal adults. Under conditions of dissociation it has no meaning: in the baby there is still no sign of the 'cyclops eye' — i.e. a functionally-unified duality. A harmonious distribution of the impulses received by the two eyes will, therefore, not be attained until connections between the representatives of homonymous retinal halves, (cerebro-retinal elements) have been established.

The inapplicability of Hering's law to the young infant emphasizes once more how necessary it is to keep our ideas about the seeing and looking of babies entirely separate from those relating to such functions in the adult.

The examination of newborns showed us that one of the first reactions to light, seen in a normal infant, was a fixation movement in which both eyes turned to the light. It may perhaps be necessary to assume that originally each eye directs itself independently towards the light, although it is possible also that use would be made chiefly of the reflex pathway for conjugate movements. It may be that the ephemeral nature of this reaction is connected with the monocular character of this fixation movement. As soon as the light stimuli have been able to act for some time, however, it is likely that a binocular fixation is brought into being by the stimulation of both eyes and that, after a short time, this can be evoked by monocular stimulation also.

The investigation also showed that the reflexes for conjugate movements predominate over the monocular reflexes, as a result of which the normal baby (at any rate during the first 3 months) always reacts with a conjugate movement to a light stimulus with sideways incidence. Finally we learned that in very young children a conjugate reaction appears earlier upon stimulation of the nasal than of the temporal half of the retina. The connections existing between the vestibular and oculomotor nuclei — connections which are phylogenetically important and are also present very early in man (5th. foetal month) — may perhaps be a contributory factor in this predominance of the conjugate ocular movements. On other grounds Rademaker & Ter Braak proved the primitive character of the conjugate movement.

Not before about the 4th. month was I able to elicit a bilateral monocular adduction movement in response to a source of light placed close in front of the eyes in the median line. Before this time such stimulation was usually not answered by any eye movement. Where such a movement does occur, the probability of a very marked tendency to squint must be taken into account.

Fusion reflexes. If this predominance of the conjugate movements were permanent, it would not be possible to fix both eyes on the same object except in very exceptional cases. To achieve this it is necessary that each eye be subjected to the required corrections. It is the monocular optomotor reflexes that have to provide this correction, thus bringing about fusion both from a motor and from a sensory point of view.

At this stage thus, the eyes are set in movement partly as a couple and partly each on its own account and dissociated movements in response to optical stimuli can, therefore, still be expected. This state of affairs may or may not be, as mentioned-before, a forerunner of a true squint.

We see thus, that by the co-operation of the reflexes for conjugate movements (which give the coarse adjustment) and the monocular optomotor reflexes (providing the necessary corrections) both eyes can be fixed precisely upon the same point. At the moment that motor fusion is established anatomically corresponding points of the retinae will be stimulated; this will often happen first in the peripheral part of the field of vision, and later in the centre, so that first the peripheral and finally also the central fusion can be achieved in both aspects. This is the foundation upon which the junction of the two eyes can subsequently be built up. All this shows up very clearly the great importance of the reflexes to conjugate ocular movements while the predominance accorded to these in process of evolution is decisive for a further normal building up of the higher optical reflexes.

In the period which follows, the influence of the developing macula will gradually make itself felt, playing an increasingly important and finally a dominant part in that interaction of reflex forces which has already been rendered so complex by the possibility of reciprocal effects.

When the convergence faculty subsequently creates the possibility of binocular vision at a short distance, it is in the last analysis this all-commanding function of the macula that co-ordinates, regulates and crowns the achievements and finally gives us the reality of binocular vision, at the same time helping to fix the position of the eyes. It is this macular function that finally, when the developing accommodation begins to influence the sharpness of images in a favourable sense and creates new possibilities in association with convergence, leads to the achievement of the highest form of optical perception, stereoscopic vision.

Optomotor reflex gaze tonus. The light stimulus which has called into being the macular dominance also makes its influence felt upon the muscular apparatus. It will not only evoke reflex contractions but will bring about changes in the tension equilibrium that has hitherto been formed by the action of idiomuscular and proprioceptive stimuli in the muscle. By way of the above-mentioned associations between the proprioceptive representations in the brain-stem and the thalamus on the one hand, and the optical cortex and the muscle nuclei on the other hand, it will evoke a changed tension-pattern — optomotor reflex gaze tonus — and will be able, moreover, to impose acute increases of this tonus as there are required for adaptation to the growing function. Here it is conceivable that this functional adaptation takes place via the higher gaze centres (in the occipital pole?).

Equipped and prepared in this way, the muscular apparatus will be capable of fulfilling high demands and, after the necessary practice, increasing sharply effected optomotor reactions will become possible, as a result of which the performance will gain in precision and the reactions in speed. This will be of particular benefit to the higher optomotor functions, while it also brings with it the possibility of attainment of a higher level.

Binocular junction. Among these higher reflexes we may certainly include the fusion reflexes. Worth felt intuitively that a disturbance of this was the key to the problem of squint, but neither to him nor to the many who came after him was an insight into the cause of the disturbed function vouchsafed.

The possibility of achieving motor fusion does indeed constitute the 'to be or not to be' of binocular single vision.

This fusion movement is indispensable for the establishment of functional junction between the 'corresponding' retinal halves and thus for the perfect co-operation of both eyes. This junction, for which I propose the name BINOCULAR JUNCTION, which seems to me a more correct and more graphic term then 'retinal correspondence', is the foundation stone of binocular single vision.

Upon this foundation other, new binocular reflexes can then be built up.

How is this binocular junction brought about? When the eyes are adjusted in the manner described above, under the influence of the reflexes for conjugate movements and the monocular reflexes, anatomically corresponding retinal elements will be simultaneously stimulated.

As a result of continually repeated stimulation there will be formed in the optical sphere of the central nervous system a junction between the representatives of these retinal elements and hence, in the course of time, between the representatives (cerebro-retinal elements) of the retinal halves of the right and left eye.

This junction will be greatly promoted by the anatomical situation of the representatives in question.

Thus the functional junction of the two eyes will take place upon the anatomically-prepared substratum.

The bridge has now been built between the two eyes. The previous dissociation has given place to an association. Impulses evoked in one eye will now be able to graft themselves upon the retinal representatives of its fellow. By virtue of this binocular junction the light stimuli which reach the optical cortex via one eye will now be able to set up optomotor impulses via the representatives of the other eye as well, so that it will now be possible for the eyes to be moved in a more or less co-

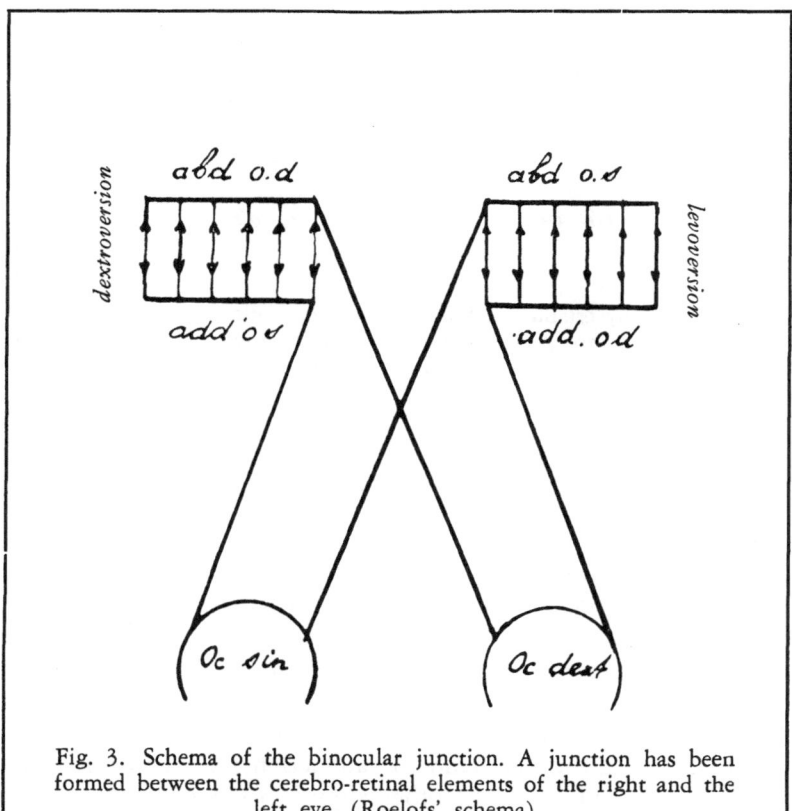

Fig. 3. Schema of the binocular junction. A junction has been formed between the cerebro-retinal elements of the right and the left eye. (Roelofs' schema)

168

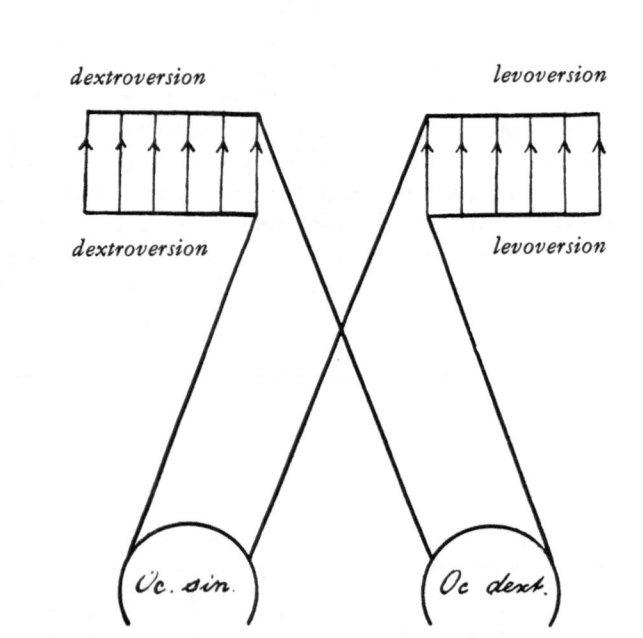

dextroversion　　　　　　　　　　　　*levoversion*

dextroversion　　　　　　　　　　　　*levoversion*

Oc. sin.　　　　　　　　　　　　*Oc dext.*

Fig. 4. Partly established binocular junction. Stimuli via the un-crossed paths are grafted onto the representatives of the nasal retinal halves. Tendency to divergent squint. (Roelofs' schema)

ordinated manner via the *monocular* reflex paths also. This co-ordination, however, is not so absolute as in the case of the reflexes to conjugate movements, it leaves enough margin to give a certain functional range. In this way a *cortical conjunction* has been established in addition to the subcortical conjunction. While, on the one hand, a stimulation of the temporal half of the retina will henceforth also be able to set up a conjugate movement, on the other hand the adduction reflexes now have the chance, by inhibition of other stimuli, to co-ordinate them-selves to a convergence reflex.

Fig. 3 of Roelofs' drawings gives the schema of the binocular junction.

This final state, however, is not achieved all at once. Under normal conditions the reflexes to conjugate ocular movements predominate over the monocular optomotor reflexes.

Of the reflexes to conjugate movements, in their turn, those reflexes reaching the eye via the crossed paths predominate over those travelling along the uncrossed paths. This we saw in very young infants. The result of this dominance of stimuli via the crossed paths will be that, as described above, the stimuli via the uncrossed paths will be grafted onto the representatives of the nasal retinal halves, as shown by Roelofs in fig. 4.

At this stage there is some danger of the development of strab. divergens, as the monocular abduction reflex now receives more stimuli than the monocular adduction reflex. This danger, however, is not great because this state of affairs has developed precisely as a result of the predominance of the reflexes to conjugate movements.

In our description of the establishment of binocular junction we took as our starting point a parallel position of the eyes.

As a rule, however, this will not be the case and the light stimulus will lead to an unequal disturbance of equilibrium in the two eyes. The binocular movement apparatus (vestibular reflex path) will then receive from each eye a different impulse towards turning in the same direction. If an equilibrium is now to be established between the cortical fields and the motor binocular centre, this is possible only if part of the impulses be conducted to the motor monocular centres (proprioceptive reflex path), which will then correct the position of each eye in such a way that an equilibrium is established between the cortical fields and the gaze centres. After these monocular corrections (fusion movements) have been applied, corresponding points will be impinged upon by identical stimuli — and this brings us round to our original starting-point.

The following diagram may help to make this clear.

A and C are cortical representatives (cerebro-retinal elements) of the right half of the left retina; B and D of the right half of the right retina. The eyes are assumed to have a divergence of 5°.

If A is struck by a light stimulus, D will be simultaneously struck by this same stimulus. From A there emerges an impulse for turning 5° to the left in conjugate movement and from D an impulse for turning 10° to the left. Since C is inhibited by

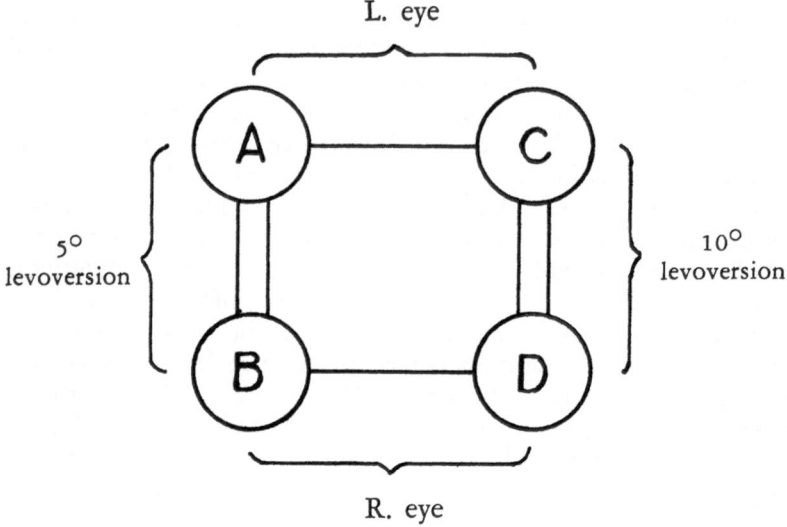

L. eye

5°
levoversion

10°
levoversion

R. eye

A and B by D, no use can be made of the binocular junction. The two impulses for turning to the left give a resultant of 7.5° turning to the left. The rest of the impulse from A is carried off via the monocular reflex path and gives an adduction of 2.5°, while the rest of the impulse from D does the same. This gives a convergence of 5°. As a result of this convergence, corresponding retinal elements are again struck by equal stimuli. In this way the fusion between the peripheral retinal elements now impinged upon by the light stimuli is established (peripheral fusion) and at the same time the way is paved for binocular junction.

By analogy with what takes place along the pyramidal and extrapyramidal paths we might imagine that the coarse binocular adjustment takes place via the labyrinthine reflex paths and the fine monocular corrections via the proprioceptive paths.

Peripheral fusion. This double mechanism of adjustment undoubtedly points to the great importance of the peripheral fusion.

This very important reflex, which was first described by Winkelman, makes the establishment of binocular junction possible. It creates, maintains and guarantees the parallel position of the eyes and, as such, constitutes the starting-point and basis for the building up of the higher binocular reflexes indispensable for binocular single vision. Finally it gives the signal for production of the chain of reflexes — of which it is also the first link — which leads to the goal of central fusion.

It will, of course, be realized that the corrections of turning applied by the peripheral fusion are in the opposite directions. The double innervation originally necessary for this will, in the course of time, be able to combine to form a single convergence impulse.

These movements of images over the retina, as produced by the peripheral fusion reflexes in the manner described, will coincide repeatedly with the subsequent stimulation of the corresponding retinal elements, which in this way themselves become reflexogenic. This creates the conditions for evocation of the following reflex — the binocular fixation reflex — which will now turn both maculae towards the light. If the macular dominance has already become a fact, the central fusion reflex will come into action, with simultaneous inhibition of the peripheral stimuli, as soon as the stimulated area approaches the territory presided over by the macular fusion reflexes.

In the same way as before, a convergence mechanism will once more come into action to direct both maculae accurately towards the light stimulus. Now the central fusion can be attained. The binocular junction, presaged by peripheral fusion, thus finds its confirmation and consolidation in central fusion.

This succession of reactions, each of which is evoked by its predecessor, must be regarded as a *chain reflex.*

Convergence. It is obvious that a convergence mechanism, such as that described in the foregoing, which is necessary for binocular single vision at short distance will not be able to make use of the reflexes for conjugate movements (vestibular reflex paths), but will be obliged to have recourse to the monocular adduction reflexes (proprioceptive reflex paths). The predominance of the adduction reflexes which is generally increasingly evident from the beginning of the 4th. month onwards, supported by a better anatomical preparation (possibly to be accounted for on phylogenetic grounds) of these paths in comparison with those of the abduction reflexes, will soon make itself manifest in a definite convergence of the visual axes in looking at near objects. This comes about at first as a result of the fact that each eye receives an impulse to adduction, due to the received retinal stimulus, and is able to respond reactively to this impulse thanks to the predominance of adduction, i.e. to the priority accorded to adduction in preference to other reactions (in this case the cortical reflex to conjugate movements.) For this purpose, thus, the impulses which would have been able to be grafted, via the cortical conjunction, on the other eye will have to be inhibited, so that the conjugate eye movement which would otherwise have been evoked in this manner does not occur.

The fact that this is a very young reflex, perhaps still hardly constituted, will facilitate this course of events.

In this way, two incompatible lateroversions are transformed into a convergence movement, with the result that the originally bilateral adduction will be able in the course of time to co-ordinate itself to a convergence reflex.

The development of the convergence reflex will possibly be aided by the synaptic junctions and associations which have formed (via the proprioceptive representatives of the left and right eye in the brain-stem and thalamus) with the cortical representatives of the retinae, and thus also between the right and left occipital poles.

It is possible to regard the convergence reflex as a special form of fusion reflex, in the sense that convergence has developed out of fusion. Like the latter, the convergence reflex also has its fusion amplitude.

True convergence, as a coupled reaction, occurs only in the primates and man, and possibly in the cat. In view of the fact that in man and apes the nucleus of Perlia pushes in between the nuclear complex of the 3rd. nerve as a new cell group in the median line at a late stage of foetal development, one is greatly tempted to assume with Brouwer that this new cell group has a function (as peripheral centre?) in convergence. This is the more attractive the more importance one attaches to the correlation between the development of this nuclear group and the position of the optical axes, a phenomenon to which Ida Mann has drawn attention.

The various higher binocular optomotor reflexes described in the foregoing can develop only on the basis of binocular junction. If, for any reason, this fails to develop, the logical conclusion is that these higher reflexes will also be unable to develop, so that the original monocular vision will be maintained at a later age; this will inevitably be expressed in the form of co-ordination disturbances. Any rational treatment of these disturbances will, therefore, as a matter of logic, have to be directed towards the establishment of this indispensable foundation of binocular linkage.

Orthophorization. Observation of the successively appearing optomotor reactions in normal babies has taught us that the first stage of this junction generally becomes established in the 2nd. half of the first 6 months. This gives support to the impression, derived from the frequency curve for strabismus, that an orthophorization process occurs in every child about the end of the first 6 months.

This orthophorization, which is a physiological development aimed at the approximately parallel direction of the visual axes — and the maintenance of this direction by peripheral fusion via binocular junction — is the outward expression of this developing binocular association. It is the product of harmonious co-operation between the reflexes to conjugate movements and the monocular reflexes. This may be regarded as the new biological equilibrium attained by means of the optical postural reflexes.

Heterophoria. This equilibrium, like any other, may be more or less easily disturbed, i.e. more or less labile. If the mutual position of the eyes, in so far as this is monocularly determined, is greatly deviated, so that the corrective monocular reflexes are barely able to achieve and maintain motor fusion, this labile equilibrium will immediately show itself in the form of heterophoria as soon as this fusion is eliminated (by covering one eye). As a rule this heterophoria will gradually decrease in the course of years. If however, the deviation is too great to be overcome by the fusion reflexes, orthophorization will not occur and thus the binocular junction will also fail to develop. The visible result of this is a permanent disturbance of co-ordination.

Physiologically the failure of binocular junction will thus be expressed in a defect of the higher optomotor reflexes. Anatomically it will be shown in an abnormal position of the eyes and owing to the absence of functional stimuli, in an impaired anatomical growth and maturation of the nerve elements in the occipital pole which constitutes the anatomical substratum upon which the physiological junction of the two eyes ought to have taken place. The observation of Max de Crinis and Berger on cytodendrogenesis and function and of Minkowski and others on myelination and function are particularly important in this respect also.

The binocular junction is thus a result of dendrite formation and myelination. Since this depends on function, binocular junction also depends on function. In contradistinction to the widespread opinion, binocular junction is, thus, not purely functional (along preformed paths) but is in fact based also upon the formation of *anatomical* connections which have developed as a result of function.

When we have spoken of fusion up to this point, we have almost always meant motor fusion. We must not imagine, however, that the monocular movements here concerned are of such magnitude that they can be easily observed. As a rule, especially in the case of peripheral fusion, there will be no question of any movement of the eyes with respect to each other; this becomes clear if we realize that binocular junction ('retinal correspondence') is not a point-to-point correspondence but is based upon the identical reaction of groups of retinal elements situated within the 'Empfindungskreis' of Panum, which are continually increasing in extent towards the periphery. For the desired result it is now sufficient if one or a few of the cells within the 'Empfindungskreis' enter into contact, via their cortical representatives, with any other point in the corresponding 'Empfindungskreis' of the other eye. The investigations of Roesch have once more definitely confirmed this while, in this connection, Zeeman has recently recalled our attention to the pseudo-problems into which the old mathematical views on the horopter question had led us, and has made a plea for a more biological approach.

Stereoscopic vision. The highest form of our vision finds its possibilities in the flexible nature of these links, in Roesch's 'zones stéréoscopiques'. When our eyes are directed to a point in visible space, this cannot simultaneously be the case in the same way for close and more remote points. It is precisely thanks to the fact that our visual axes are not directed strictly parallel on all these points, so that in this respect we physiologically always 'squint' slightly, that we are able to enjoy this highest form of human optical perception, three-dimensional vision.

II POTHOLOGICAL SECTION.

As we saw in the first section of this chapter, the optomotor reflexes become manifest in a definite succession in the normal infant.

In a rather large number of cases, however, deviations from this normal system of development occur and lead to certain consequences as regards the position of the eyes.

In these cases we find, sometimes very soon after birth but usually in the course of the first few months, that the monocular adduction reflexes predominate over the reflexes to conjugate movements and over the abduction reflexes.

This predominance of the adduction reflexes may be due partly to a better anatomical preparation (perhaps to be accounted for on phylogenetic grounds) of the paths in connection with convergence, in addition to which, however, there would then have to be a retardation in the normal development of the pathways for the conjugated reflexes.

This opinion might be supported by the fact that the stimuli via the uncrossed paths have precedence over those via the crossed paths.

A disturbance of myelination might play a part in this deviation from the normal chronological order.

It is, however, much more likely that this condition is not a disturbed chronological order of normal development but a retarded development of the preparation for the optomotor reflexes as a whole, while the chronological order of events remains normal. This order, then, is important.

There are several reasons for believing it to be as follows:

(1) monocular adduction;

(2) conjugate movement via the crossed paths;

(3) conjugate movement via the uncrossed paths;

(4) monocular abduction.

Everything will now depend upon the stage in this series

at which birth intervenes. If birth takes place at a moment at which none of the optomotor reflex paths is sufficiently prepared, we get the picture with which we have become familiar in the case-histories of myelogenesis retardata in Chap. VI. If at birth only the optomotor reflex paths for monocular adduction are ready for use, this pronounced predominance of the adduction reflexes will show itself in a marked adduction position — often of both eyes — which is the prototype of strabismus convergens.

If the preparation of the optomotor reflex paths at the time of birth has already advanced considerably further, so that the predominance of the adduction reflexes has been weakened, a rapid spontaneous disappearance of the strabismus convergens may be expected.

If, finally the optomotor reflex paths for conjugate movements are already sufficiently prepared at the moment of birth, the chance of a convergent squint is small, although the persistence of the reflexes to monocular adduction still continues for some time to make this possible.

We still have to explain why in normal babies the monocular adduction does not become evident until about the end of the 3rd. month, whereas in squinting babies it is one of the first reactions detectable. This must be accounted for by the predominance normally held by the conjugated reflex. Thus, the conjugated reflex keeps the monocular one in check. In squinting children this is not the case because the reflexes to conjugate movements are still undeveloped or insufficiently developed.

From all this it follows that in older normal children this braking effect can only be abolished by inhibition of the conjugated reflexes. At this time, thus, the child must already be able to *see* and associative links must have been formed. Here we have the explanation of the late manifestation of the adduction reflexes in normal infants.

It is conceivable that in this way strabismus can still develop at a later age, at a time when this *seeing* has acquired greater practical significance, especially where the optomotor reflexes to monocular abduction are poorly developed and are unable to exert a counterbalancing effect.

As regards the clinical form in which strabismus convergens may appear, the following remarks may be made.

The prototype, with a very strong bilateral predominance of

the adduction reflexes over the other reactions, gave us the picture of strab. bilateralis. Apparently the predominance was here practically equal on both sides, while the abduction was still very weak or absent. This picture is sometimes seen clinically in very young children, especially in those with severe retardations. As we saw in Chap. III, this picture can often be reconstructed in the dark-room.

The fact that this has to be done in the dark-room is probably due to the fact that stimuli from the periphery — which habitually fix the position of the eyes — are then much less active.

If the conditions are the same for both eyes a choice — which cannot be further defined — will decide which of them is to become the leading, dominating eye .As is well-known, nearly 98 % of human beings have a preference for one eye. If the eyes are unequal, the better one will generally be chosen as leader. The study of uniovular twins has shown that external factors, not situated in the eye itself, may also participate in this choice. In one of the writer's cases (published by Waardenburg) [1]) a discordant ptosis decided the mirror-image concordance of strabismus unilateralis. In non-twins, an unequal development of the abduction reflexes can, of course, also decide the choice.

Among the factors situated in the eye itself, the refraction difference occupies only a subordinate position, since in the overwhelming majority of cases the refraction in young squinting children is the same on both sides (see Chapter III).

We may thus expect that where there is no definite preference for one eye strabismus alternans will develop, while in the other cases strabismus unilateralis will appear.

It is understandable that under altered circumstances the one form of squint may change into the other, as clinical experience confirms. Authorities on genetics are also of the opinion that the two forms are merely expressions of the same disease. From this view of the clinical pictures, the secondary character of the amblyopia can be readily deduced.

So far we have more or less tacitly assumed that the cause of the predominance of the adduction reflexes has to be sought in a central direction. One might consider, however, that the retina ought also to play some part in the pronounced motor predomin-

---

[1]) Documenta Ophthalmologica p. 172, 1950. Twin research in ophthalmology.

ance after stimulation of the temporal retinal halves, the more so as many authors (Harms and others) consider retinal inhibition to be an important element in the problem of amblyopia.

In order to decide this point the aid of the electroretinogram was called in; this not only gave exactly the same result with the amblyopic as with the good eye but also failed to show any significant difference between the tracings (ERG) obtained after stimulation first of the temporal and then of the nasal part of the retina, either in the good or in the amblyopic eye.

On these grounds, therefore, we may safely assume that the retinal function is normal in both eyes and that the cause of the adduction predominance, and hence of the strabismus convergens, cannot lie in the retina.

If objective proof were now forthcoming that the affection is situated in the central nervous system it would be possible — after it had been shown by means of optical and vestibular stimulation that the eye was capable of normal movement in all directions — to consider the existence of a peripheral cause of strabismus as satisfactorily disproved, except in a few rare cases of peripheral anatomical abnormalities in the muscles and ligaments or in the orbits.

The electroencephalogram (EEG) undoubtedly offers the possibility of such an objective proof. Attention has recently been drawn, especially in America, to the existence of occipital involvement in ocular affections, including muscular paresis of central origin.

A recent investigation, by Levinson and Stillerman, on the correlation between EEG changes and eye disorders showed that in a group of 915 children under 16 years of age 71 % of cases of paralysis of central origin with eye disorders had occipital involvement, while the latter was found in only 23 % of cases with eye disorders only and in 0.6 % of completely normal children. Their conclusion was that an EEG should be taken in all cases of any kind of eye disorders in children. (EEG Journal, May 1950)

Since the nature of the EEG pattern is partly dependent on the number of functioning nerve elements, it seems to me not a priori excluded that a functional disturbance in the occipital pole might manifest itself in slight deviations from the normal pattern for the age in question. A comparison of the EEG of the occipital pole at rest and with illumination would further provide information as to the functioning of the centripetal paths, while a simultaneous recording of the ERG (electroretino-gram) would provide a number of other interesting obser-vations.

With the aid of optokinetic nystagmus I attempted also to get some impression as to the functioning of the various path-ways in young squinting children; I found this reaction to be absent in the first 3 to 4 months. This investigation is still incomplete and for that reason, as well as the necessity for keeping the present study within certain limits, I shall not go into it further on the present occasion.

Thanks to the kind co-operation of Dr. J. Droogleever Fortuyn, attached to the 'Herseninstituut' (Brain Institute) in Amsterdam, I had the opportunity of gaining for myself a preliminary impres-sion of the possibilities that EEG examination of the squinting child may provide. At the present stage of the investigation it is much too early to pronounce any judgement but one can safely say that this seems rather promising.

What are now the consequences of the condition resulting from the predominance of the adductors?

First of all there is the direct consequence in the form of a deviation (in adduction) of one or both visual axes. The size of the angle of strabismus thus produced may vary greatly according to circumstances. In young children there is no quest-ion of a fixed angle of squint. The deviation may vary from 0 tot 30°, so that its average will be about 15°. Measurements of this angle in young children are of little or no value.

Upon horizontal movement on optical stimuli in the field of gaze a limited abduction of the squinting eye will be seen; this is the result of the adduction prevalence, which may or may not be reinforced by an absence or low degree of development of abduction.

This accounts for the temporal limitation of the 'horizontale Bewegungsstrecke' (horizontal movement range) of Nordlöw and others.

An obvious indirect consequence will be that the central epicritical vision of the deviating eye will have its development prevented or impeded so that the condition of amblyopia will result. This interruption of function will make its repercussions felt on the anatomical development of the corresponding ele-ments of the brain, which will mean that the state of develop-ment existing at the time of appearance of the amblyopia will

180

remain. As we have ascertained clinically, however, it is always possible during early childhood to arouse and improve the central visual acuity by means of training of the relevant optomotor reflexes, the attainment of a good central visual acuity being the general rule.

All this time we have been arguing from the premise that the deviation of the visual axes produced by the adductor predominance is so large that the fusion reflexes are incapable of bringing these into a parallel position. The chief and the most far-reaching consequence of an anomalous position of the eyes is, however, the impossibility of attainment of binocular junction.

If the fusion reflexes have failed to bring the eyes into a parallel position, in the manner described in the first part of this chapter, the opportunity for repeated simultaneous stimulation of corresponding points in the homonymous retinal halves will be lacking. Thus the peripheral fusion will not be achieved and the result of this is that the possibility of establishment of a central linkage (junction) between the representatives of the corresponding retinal halves of the two eyes is also excluded.

In squint, thus, the primitive dissociated position of the eyes will be maintained. This fact has been clinically confirmed by the possibility of evocation of monocular reactions and the clearly evident dissociation of the eyes in horizontal movements in the binocular horizontal field of gaze.

Strabismus is called into being by the predominance of the monocular adduction reflexes over the reflexes to conjugate movements and the abduction reflexes, and is consolidated by the incapacity of the fusion reflexes to correct the deviation once established.

This shows the correctness not only of Worth's intuition but still more of the extensive and important investigations of Roelofs, who was repeatedly led to the conclusion that the cause of squint must lie in an insufficiency of the fusion reflexes. [1])

It is logical that the building up of binocular junction should be correlated with the development of the optomotor reflexes and of the optical cortex. As we have already seen, the first periode of establishing falls within the first 6 months of life. It proceeds to completion upon a suitable prepared anatomical substratum.

[1]) C. Otto Roelofs. Die Fusionsbewegung der Augen. Arch. f. Augenheilk. 97, 229, 1926.

According to Kleist, who based his opinion upon the results of the topographical anatomical studies of Cramer and Minkowski, the anatomical substratum of these junction probably lies in the double granular layers of the IVth. cortical layer; this opinion is supported by the results of Volkmann's comparative anatomical research. Clinical data on injuries to the occipital cortex also point in the same direction.

In connection with the correlation between function, and cytodendrogenesis and myelination (Berger, M. de Crinis, Monakow, Minkowski) — tersely expressed by C.U. Ariëns Kappers in the words 'functions dominate forms', it goes without saying that the process of building up this junction will take place gradually and that all kinds of individual differences will be seen in it. Transition stages between the normal presence and the complete absence of linkage are to be expected, and can also frequently be objectively demonstrated upon clinical examination. They lead to more or less severe co-ordination disturbances which may be manifested in the form of a periodical or continuous strabismus.

With the aid of the schemas worked out for this purpose by Roelofs we shall now try to trace the various possibilities which may present themselves in the formation of cortical junction between the representatives of homonymous retinal halves and also the consequences of such possibilities. Here we shall quote frequently from the text appended by Roelofs to his schemas. We shall discuss in turn:

(a) complete absence of binocular junction;

(b) defective development of binocular junction;

(c) abnormal binocular junction;

(d) presumably abnormal binocular junction.

(a) *Complete absence of binocular junction.*

If birth occurs at a moment at which only the reflex paths for the monocular adduction reflexes have reached a sufficient stage of preparedness, this will very soon show itself in a marked bilateral adduction position of the eyes (see Chapter IV). In such cases the lay person speaks of a congenital squint. This, of course, is not correct, since a *true* squinting position can only have developed after optical stimuli have made themselves felt.

In these cases of marked bilateral adduction appearing at a very

182

early age, it is obvious that the fusion reflexes (among which in this case the monocular abduction reflexes must also be included) will not be capable of directing the eyes in such a way that corresponding points in the homonymous retinal halves can be simultaneously stimulated. If this clinical form passes over, as it always does, into analternating or unilateral strabismus, the pronounced deviation of the visual axes will mostly, especially in the alternating form, provide too heavy a task for the fusion reflexes, which will show themselves unable to cope with is.

The immediate consequence of this is a continuation of the primitive monocular vision, i.e. a complete maintenance of the dissociation. It is quite understandable that these cases without

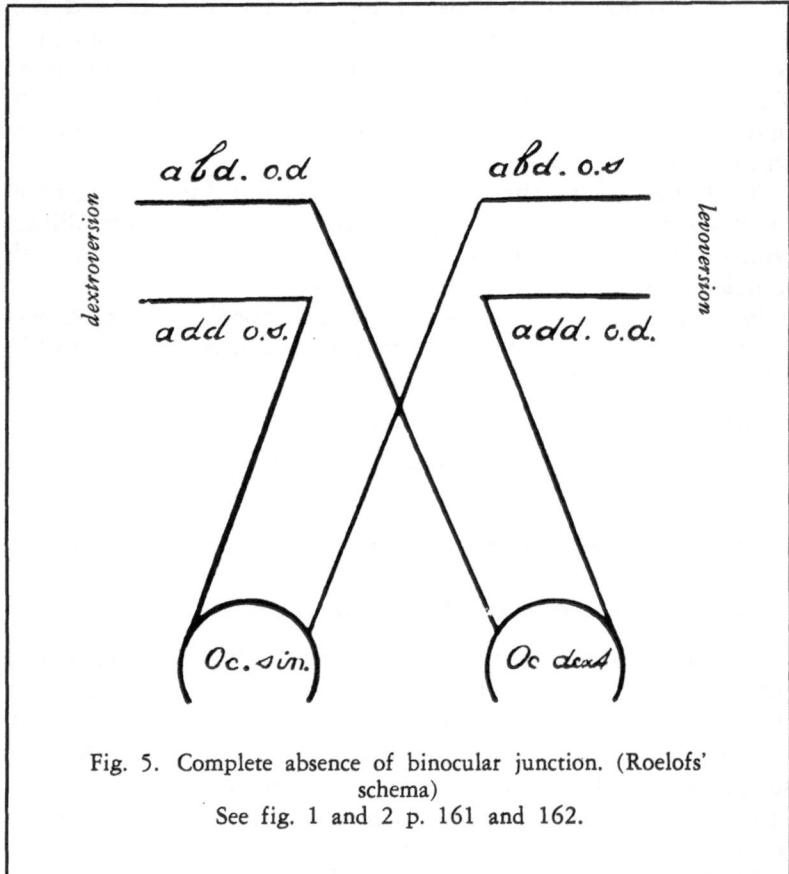

Fig. 5. Complete absence of binocular junction. (Roelofs' schema)
See fig. 1 and 2 p. 161 and 162.

any junction whatever will resist practically all the usual forms of conservative treatment of squint as long as the adduction maintains its marked predominance and the abduction reflexes do not achieve a sufficient counterbalancing effect.

Fig. 5 gives the schema for this condition, in which the reflexes for both monocular and conjugate movements travel exclusively over the pathways for each eye separately.

(b) *Defective development of binocular junction.*
Here there are various possibilities.

One must bear in mind here that for the junction to function properly it is necessary that the stimuli be able to cross the bridge without hindrance in both directions.
If the deviation of the visual axes is such that, under certain

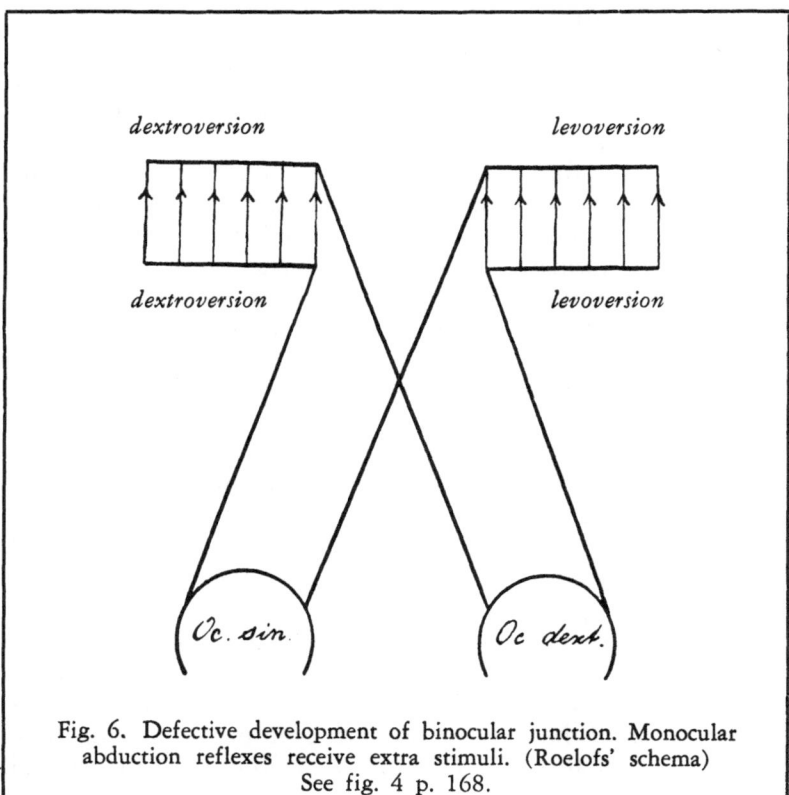

Fig. 6. Defective development of binocular junction. Monocular abduction reflexes receive extra stimuli. (Roelofs' schema) See fig. 4 p. 168.

conditions, it is just possible to achieve peripheral fusion, the junction will be established only in a gradual manner; this will favour the occurrence of periodical strabismus.

Under certain conditions a junction may have been partially established but may be lost again. This may happen if the following reflex, as shown in the development schema (see p. 175), does not manifest itself at the proper time, so that any predominance present will be maintained too long.

Under normal conditions, as we saw in the first part of this chapter, the reflexes to conjugate movements predominate over the monocular optomotor reflexes. In the reflexes to conjugate movements, in their turn, those reaching the eye via crossed paths predominate over those making use of the uncrossed paths.

The fact that the binocular junction has not yet been established, makes this understandable.

As a result of this predominance of stimuli via the crossed paths the stimuli via the uncrossed paths will be temporarily grafted onto the representatives of the nasal retinal halves, as shown in fig. 6.

At this stage thus, there is a danger that strab. divergens may develop, because now the monocular abduction reflexes receive more stimuli than the monocular adduction reflexes.

If now, as normally occurs, the adduction reflexes come to the fore in time, this tendency to divergence will be easily and promptly corrected and the binocular junction will develop normally.

If, however, the adduction reflexes fail to appear in time, and if in the meantime the divergence has attained such a degree that the adduction reflexes are doomed to failure in their efforts to correct it, the divergence will become manifest and develop into a continuous divergent squint. We are of the opinion that the origin of many cases of divergent squint is to be explained in this way. An already formed binocular junction will be lost again in this way.

The conjugate reflexes as a reaction to a stimulation of the temporal retinal halves — i.e. via the cortical conjunction — will now not be established, which means that possibility of a differentiation of the convergence reflex out of two incompatible lateroversions is also lost. This can objectively be demonstrated upon clinical examination.

Dissociated monocular movements remain possible in these cases. Everything will depend here upon the moment at which,

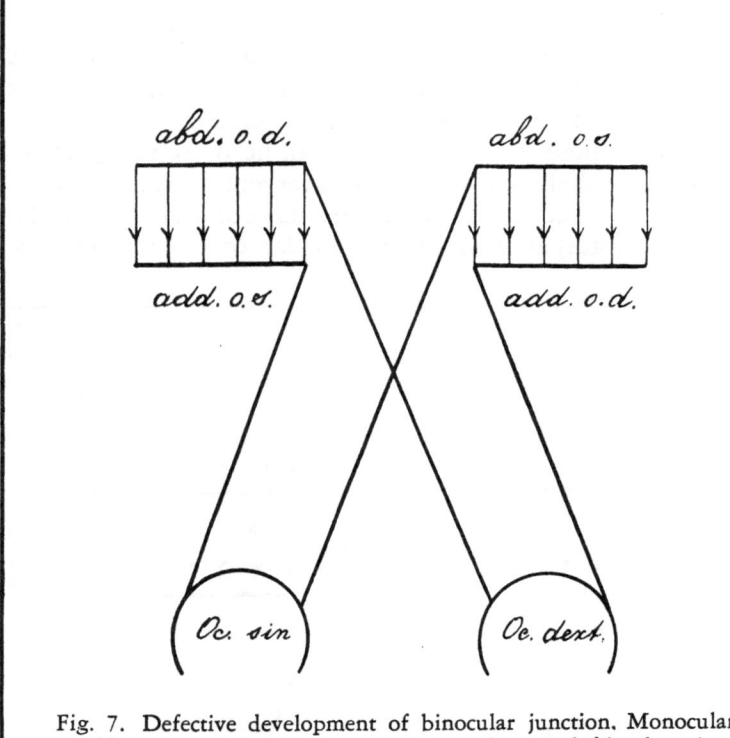

Fig. 7. Defective development of binocular junction. Monocular adduction reflexes receive extra stimuli. (Roelofs' schema)

and also the extent to which the adduction reflexes may perhaps still develop. Individual differences will undoubtedly be found here.

We can imagine, for instance, that for some time the adduction reflexes still manage to bring about peripheral fusion, but that their predominance never becomes sufficient to permit them to co-ordinate themselves in the course of time to a convergence reaction.

These differences in cases of strabismus divergens can be very satisfactorily observed clinically. We see all possible transitions between a normal convergence reflex and a bilateral adduction which can hardly be maintained even momentarily and which is usually very unequal in the two eyes. In many cases one eye does not even take part in the reaction at all.

In accordance with its mode of origin, strabismus divergens does not as a rule develop early. Cases in which it appears rather late (3rd. to 5th. year) are not uncommon.

Since we propose in this study to confine ourselves to strab. convergens, we shall not discuss the divergent anomaly further. We have mentioned it simply because its origin also has to be sought in a disturbance of the monocular optomotor reflexes.

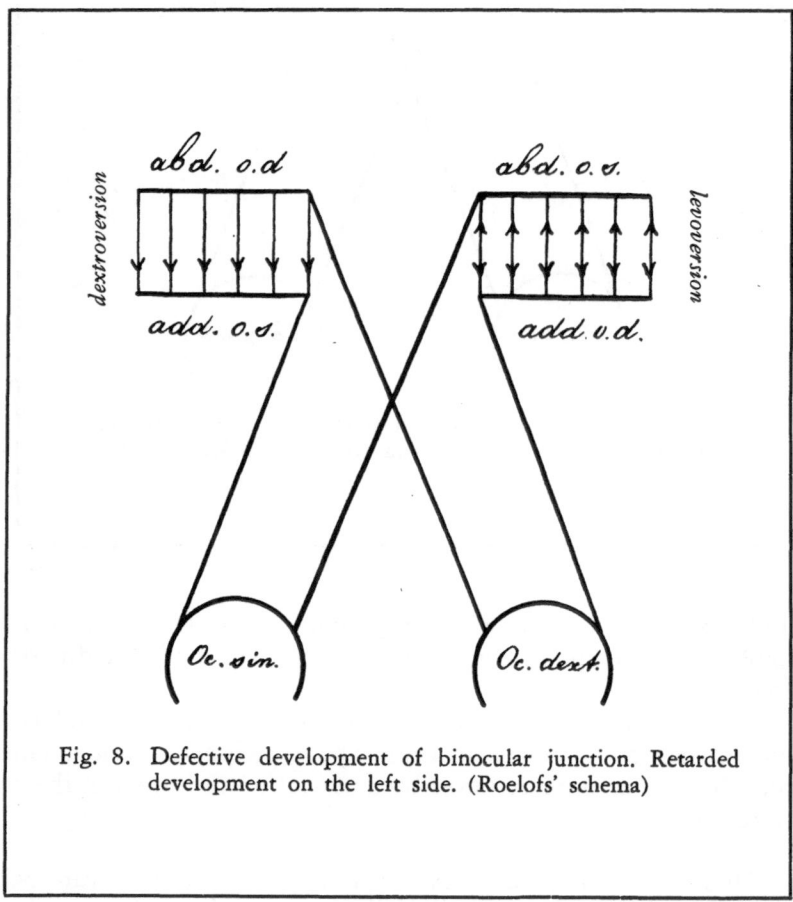

Fig. 8. Defective development of binocular junction. Retarded development on the left side. (Roelofs' schema)

In the foregoing we were concerned with the consequences

of a retarded manifestation of the monocular adduction reflexes. A quite different state of affairs results, however, if these reflexes, on the contrary predominate at a given moment over the reflexes to conjugate movements. As a consequence of this predominance we then get a grafting of the stimuli from the nasal retinal halves on to the representatives of the temporal retinal halves. In this way the danger of development of strab. convergens becomes very great, because the monocular adduction reflexes now receive extra stimuli. It will now depend on the degree and duration of the predominance of these adduction reflexes whether a convergent squint, continuous or otherwise, will develop or not. If the fusion reflexes can just manage to produce a parallel position of the eyes and to maintain this, the result will be only a more or less pronounced heterophoria (esophoria). If, under the influence of opposing factors, this can be achieved only intermittently, a periodic squint will appear. If this becomes continuous,, the binocular junction, which is still only partly established, may be lost again. Obviously, the angle of squint in the cases mentioned-above will originally be highly variable; this tallies with our clinical observations.

A squint originating in the manner described above will, as a rule, not be able to appear before the 2nd. half of the first 6 months. In view of the observations on the age of onset of squint (Chapter II), which show the highest frequency of appearance to fall in the first 6 months, we may assume that the majority of cases of squint originate in the manner described here. A beginning of binocular linking had already formed, but at a later stage this was lost. Fig. 7 of the schemas depicts this mode of origin of strabismus convergens.

Still another form of defective binocular junction presents itself when this process runs normally in one occipital pole but is retarded in the other. (see fig. 8). Such cases can be clinically observed and diagnosed. The following paragraph, from a private communication from Roelofs who examined such a case, is quoted with his kind permission:

'In this case a slight amblyopia of the right eye is to be expected, this being connected with the partial grafting of the stimuli, onto the representatives of the left eye. In the dark the eyes are more or less correctly directed. In the light they are also usually correctly directed. All eye movements are correctly performed. When the good eye (left) is covered the

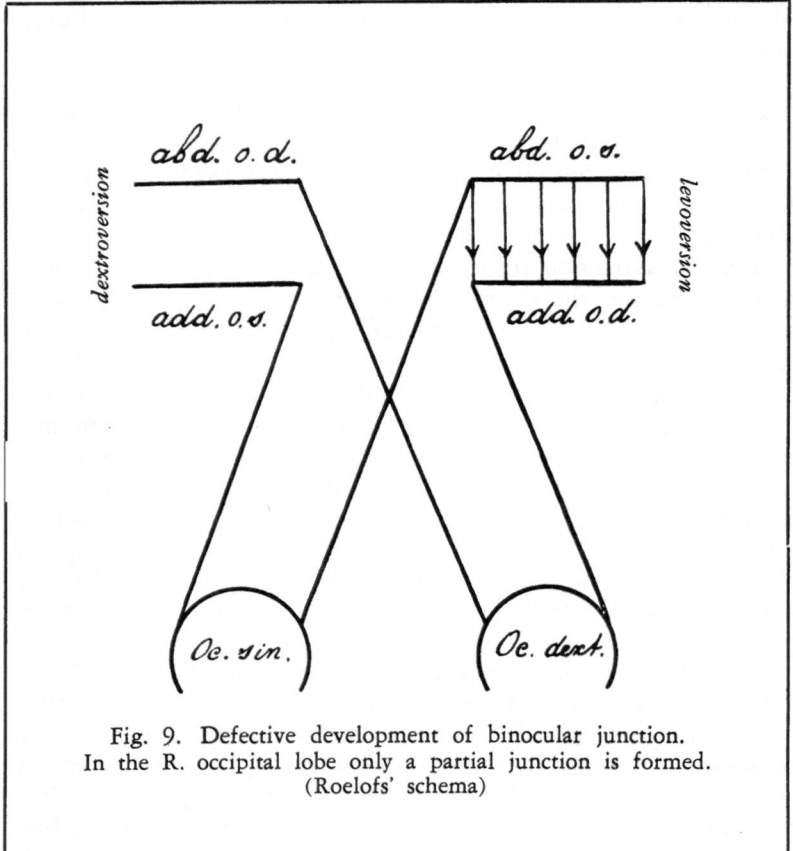

Fig. 9. Defective development of binocular junction.
In the R. occipital lobe only a partial junction is formed.
(Roelofs' schema)

eyes still remain approximately correctly directed. When the slightly amblyopic (right) eye is covered, however, this eye slips into a more or less pronounced adduction. The explanation is obvious. The light stimuli received by the good left eye give an increased gaze tonus both over the paths for conjugate movements and over those for monocular movements. On account of the only partially formed binocular junction in the one cerebral hemisphere, the right eye receives stimuli for increased adduction tonus but not for increased abduction tonus. When the right eye is uncovered again, however, the abduction tonus of this eye is restored again by the light stimuli.'

Finally, the case is theoretically conceivable that a binocular junction which is only partial may have formed in one occipital pole, while in the other pole it is entirely absent. Fig. 9 shows the schema for this. Via the partial link only a grafting of stimuli from the crossed path (left eye) onto the end stations of the uncrossed path (right eye) can take place.

Since in our schema the grafting of the left eye takes place upon the right eye, the left eye will be more or less amblyopic.

If this patient now looks with his good right eye, the left eye will not receive, via the right one, any stimulus to monocular movement. If, however, the right eye is covered and the patient made to fix with the bad left eye, the right eye receives via the crossed path a stimulus to adduction.

In contradistinction to the previous case, thus we see here the good eye falling into an adduction position when covered, whereas covering of the amblyopic eye causes no change in position.

Although such a case is certainly a possibility in the formation of binocular junction, at an early stage, I have never so far managed to find one. It is certainly desirable to continue the search.

(c) *Abnormal binocular junction.*

In the literature the name abnormal 'correspondence' for this condition has gained a foothold. This term is, however, incorrect. The correspondence of the retinae is based on an anatomical foundation that is always present and always the same. This correspondence cannot be abnormal. What can happen is that in squint an abnormal binocular junction may form upon this normal foundation of anatomical correspondence, i.e. a junction between *non*-corresponding points. In the interests of accurate terminology it would be very desirable to replace the term 'correspondence' for the physiological linking of the homonymous retinal halves by 'binocular junction'. This may be normal or abnormal.

An abnormal binocular junction may develop with a constant angle of squint, when the normal junction has never been able to develop or has been lost. It may thus occur in some cases of strab. convergens alternans or in old cases of unilateral squint. In the abnormal binocular junction we may see a compensatory reaction aiming at some degree of binocular vision without diplopia. It is always secondary to squint and can therefore never be its cause.

An important point is that abnormal binocular junction is associated with a peculiar kind of optical localization. If the examination of localization compels one to conclude that there is binocular vision without diplopia, the presence of an abnormal binocular junction is practically certain. Caution is required, however, in the assumption that such a junction exists. This anomaly will not, in the nature of things, be found in very young children. Fig. 10 gives a schema of this junction for strabismus alternans.

In cases of abnormal binocular junction with unilateral squint the state of affairs is again somewhat different, because here the abnormal junction will not be the same on both sides. This

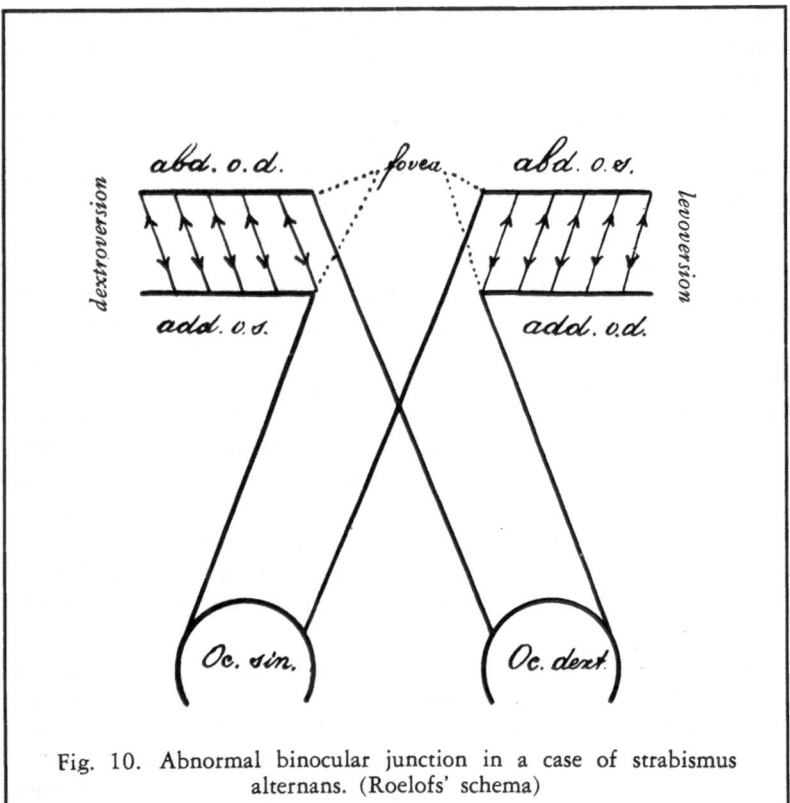

Fig. 10. Abnormal binocular junction in a case of strabismus alternans. (Roelofs' schema)

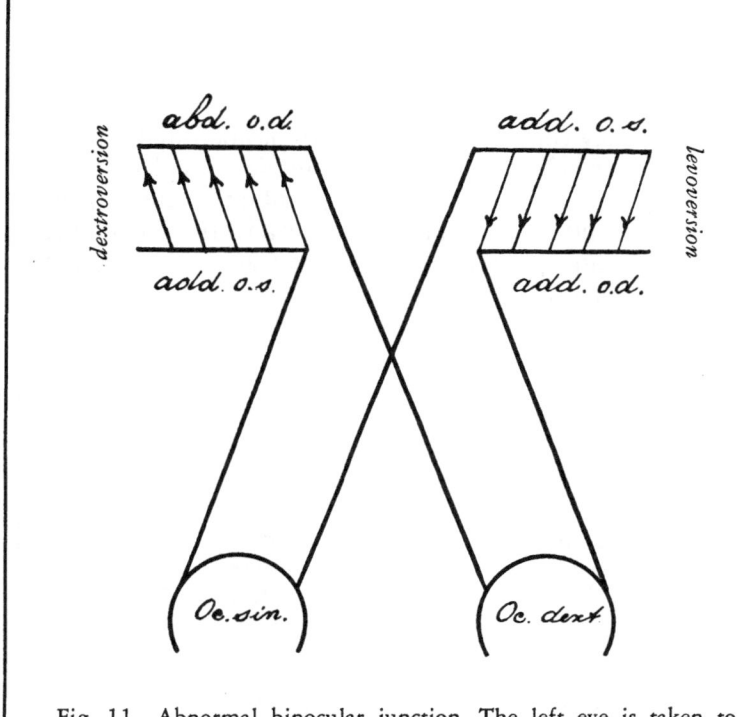

Fig. 11. Abnormal binocular junction. The left eye is taken to be the inferior one. (Roelofs' schema)

inequality, i.e. the fact that the binocular junction on one side has reached a different stage of completion from that on the other, with all the consequences thereof, has already been mentioned more than once. For this reason it seems better not to speak of *the* binocular junction but to distinguish between right junction and left junction and to judge these separately.

If an abnormal junction exists in an early case of strabismus unilateralis, the abnormal junction is not the same on both sides. If the left eye is taken to be the inferior one, the stimuli from this inhibited, amblyopic eye will graft themselves onto the representatives from the other eye and will send their motor impulses via these to the gaze centres; in this way they will also localize via

these representatives of the other eye (Roelofs). See schema 11
The optical localization will not be further discussed here.

(d) *Presumably abnormal junction.*

Finally we may consider the possibility of abnormal binocular junction in cases which suggest the conditions depicted in the schemas in figs. 7, 8 and 9. Various possibilities are shown in schemas 12, 13 and 14, which may serve as a working hypothesis. Here we have an approximately constant angle of strabismus with an amblyopic eye.

Fig. 12 assumes a predominance of stimuli from the temporal

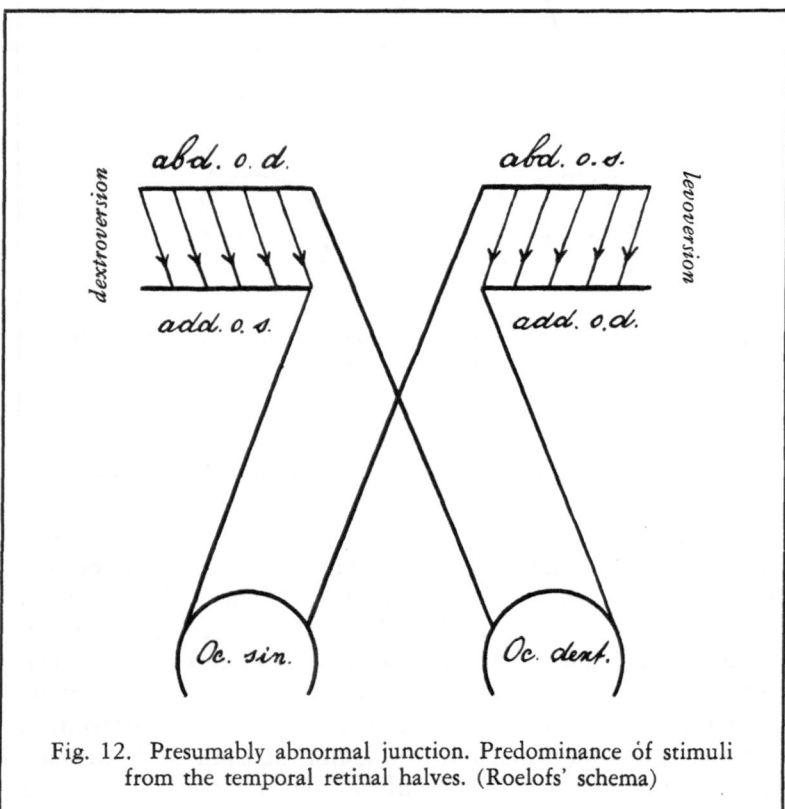

Fig. 12. Presumably abnormal junction. Predominance of stimuli
from the temporal retinal halves. (Roelofs' schema)

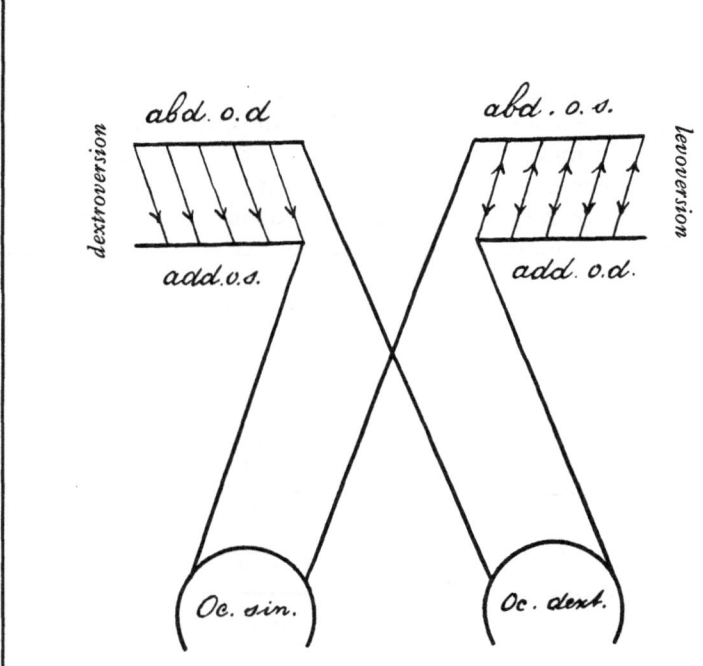

Fig. 13. Presumably abnormal junction. The right eye is the amblyopic one. (Roelofs' schema)

retinal halves in the formation of the abnormal junction. Upon covering of either eye, the angle of strabismus in the covered eye might increase somewhat. This variation of the angle of strabismus, however, makes it less probable that abnormal binocular junction will take place in these circumstances.

The condition in fig. 13 is reminiscent of that shown in fig. 8. The right eye is here the inhibited, amblyopic eye. Upon covering of this less good eye, an increase of its angle of strabismus may be expected.

In a case such as that illustrated in fig. 14 the abnormal junction is very loose. The condition is reminiscent of that corresponding to fig. 9. The left eye is here taken to be the inferior one, which is partly grafted onto representatives of the right eye. Covering of the better right eye might here lead to an adduction of this eye, owing to the stimuli which it receives via its cortical representatives from the left eye.

The cases and possibilities associated with abnormal binocular junction and the consequences of this for optical localization have been mentioned here for the sake of completeness, and also because they impose special demands in connection with treatment. As we have already remarked, abnormal binocular junction does not enter into consideration as a cause of squint.

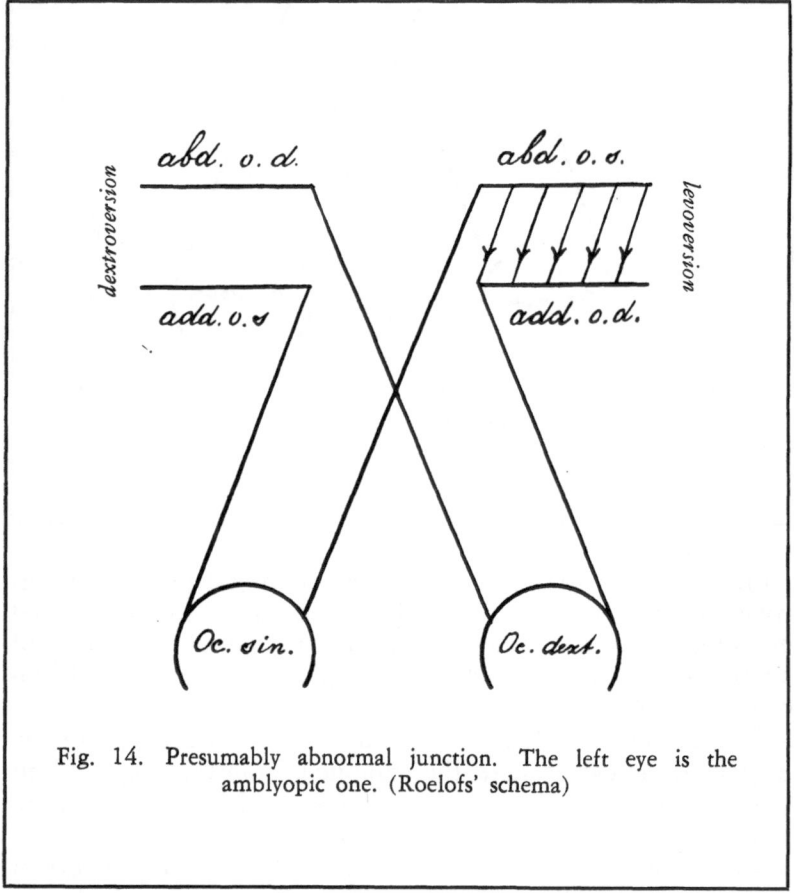

Fig. 14. Presumably abnormal junction. The left eye is the amblyopic one. (Roelofs' schema)

During the process of orthophorization, all kinds of factors, both endogenous and exogenous, may impede the establishment of binocular junction and thus favour squint. If, notwithstanding these disturbing factors, junction does to some degree take place it will be rather labile and may be lost again under the influence of circumstances arising at a later date.

In addition to primary squint, which has developed exclusively under the influence of the predominance of the monocular adduction reflexes, we also distinguish secondary squint, in which secondary factors have played the decisive part. These are not causal but only contributory factors.

The occurrence of secondary squint is responsible for the more gradually descending course, interrupted by a few rises, which we see in the frequency curve after the first 6 months.

Among the factors promoting squint we must include disturbances or abnormalities of the general physical and mental condition and development, serious nutritional disturbances, infections (especially cerebral and broncho-pneumonial), anomalies of the skull or of the muscles and ligaments of the eye and finally abnormalities of the eye itself, such as congenital cataract, clouding of other media, ocular malformations, retinal affections, aniseiconia and errors of refraction.

In this connection it may be asked how the fact that the overwhelming majority of strabismus convergens patients are hypermetropic can be accounted for. This correlation between hypermetropia and squint is the more interesting in view of the fact that in this study we have been forced to conclude that the refraction condition has no influence of any importance on the development of primary squint.

Starting from Hartridge's suggestion that growth disturbances may be a factor in the establishment of refraction conditions (as we have also assumed to be the case for squint) we might consider hypermetropia and squint to be two consequences of a retardation which is possibly due to hormonal disturbances.

The following explanation, however, seems more probable. Since the overwhelming majority of children are a few dioptres hypermetropic at birth, it is inevitable that among the squinting babies there must be a large proportion of hypermetropes. To these we must add the cases in which the squint develops secondarily at a later period, under the influence of the association between accommodation and convergence at the time when close vision begins to assume a more important rôle.

However, it still remains noticeable that emmetropes are practically never found among squinting individuals, although this refraction condition is believed, in accordance with recent observations (Stenström), to occur in at least $^1/_3$ of humanity.

Thus we might re-phrase our question and ask: why does not emmetropization occur in patients with strab. convergens?

In 1933 Sorsby, in an investigation of the refraction in school-children, ascertained anew that a change in the refraction condition takes place during childhood, amounting on an average to 3 D. According to the values found by the present writer in young children with strabismus convergens (most of them fall in the refraction group of $1\frac{1}{2}$—4 D hypermetropia) it might appear that many of these children had indeed become emmetropic.

The fact that this did not occur may be ascribed to the fact that the predominance of the monocular adduction reflexes interferes with the establishment of a high visual acuity. The last-named depends on the degree of training of the central monocular optomotor reflexes (see Chap. VIII Amblyopia). The more sharply the optomotor stimuli for each retinal element are adjusted the higher can the visual acuity become. A high visual acuity promotes emmetropization. In the case of strab. convergens its achievement is rendered difficult by the disturbance of the monocular optomotor reflexes. Thus, the same disturbance which leads to the faulty position of the eyes is also responsible for the fact that the natural correction of hypermetropia fails to occur in strabismus cases.

From the introduction to this study we have seen that all above-mentioned contributory causes, either singly or in combination, have for many years been regarded as causal factors in squint.

The true cause must be sought, as I believe I have been able to show, in a disturbance in the manifestation of the optomotor reflexes, expressing itself in the predominance of the monocular adduction reflexes over the reflexes to conjugate movements and the abduction reflexes, during a certain stage of development.

It is, thus, not an anatomical, mechanical or pathological cause that lies at the root of strabismus but a slight deviation from the evolutionally-determined time at which the various optomotor reflexes normally manifest themselves. As a result of the failure of binocular junction to take place under these circumstances, Nature's goal of the ideal association of the two eyes to a unity at a higher level cannot be reached, and the outward manifestation of this is an abnormal position of the eyes.

This absence of binocular junction as the pre-eminent consequence of the adduction dominance leads to important consequences. All the higher binocular reflexes which are superposed on this junction, such as the cortical reflex to conjugate movements and the chain of binocular fixation reflexes (among which we may probably include convergence as well) which begins with peripheral fusion, is now unable to develop.

In this way binocular single vision is deprived of its foundations and its means of existence.

The Anglo-American literature rather frequently mentions a disturbance of these binocular reflexes as a probable primary cause of squint. As we have been able to show, however, this disturbance is secondary in character.

The primary cause of squint must be sought not in a disturbance of these higher binocular reflexes but in a departure from normal in the manifestation of the lower, monocular optomotor reflexes.

At this point we may well ask what is then responsible for this predominance of certain optomotor reflexes.

The time at which the affection becomes clearly visible already gives us a valuable hint. It has been ascertained by clinical and statistical investigation that this onset of squint occurs very early, as a rule even before the age of 6 months. The observations have also taught us that this strabismus has to be regarded as the consequence of a disturbed stage in the physiological development of the optomotor reflexes, a process which normally leads to orthophorization and junction of the two eyes and which every child thus has to pass through in its first period of growth. We have also seen that the anomaly has a strong tendency to spontaneous recovery. All this makes it clear that we are concerned here with a growth disturbance, the consequences of which disappear spontaneously upon further growth, i.e. with a retardation. Further, everything points to the conclusion that this retardation is localized centrally, and more particularly in the occipital pole of the brain. On the grounds of observations on children with 'papilla grisea' and similar disturbances of the optical system, in addition to the results of the study of postnatal growth of the brain, the logical thing is to think in the first place of disturbances of myelination. (myelogenesis retardata).

A delayed myelination, in the wider sense of the term as previously described (Chapter VI), of certain parts of the optical pathsways would then be responsible for the disturbance in the manifestation of the optomotor reflexes. This retardation, or its immediate cause, would — as the sole cause of squint — also account for the simple dominant hereditary transmission of the affection, so that a genetic problem might be solved in this way too.

Obviously, this retardation must in its turn have a cause. Both the nature of the disturbance and its high degree of hereditary occurrence point in the direction of a hormonal influence. It is conceivable that a disturbance in the foetal or postnatal endocrine system of the infant might be responsible for the retardation. This view may derive support from the present discussions on the adrenocorticotropic hormonie (A.C.T.H.) and cortisone in relation to some diseases in prematures and infants, in addition to Hartridge's opinion that hormonal influences on the growth of the eye are probably involved in the development of refractions.

This view of the fundamental cause of strabismus might open up prospects of its rational treatment.

SOME REMARKS ON THE TREATMENT OF SQUINT.

A study such as this would not be complete without a few remarks on the therapeutic aspects, at any rate in so far as these follow directly from the results of the investigation. At present we shall confine ourselves to indicating a few guiding lines, leaving more detailed consideration to a future occasion.

The final aim of treatment must be to preserve or attain a normal visual acuity in both eyes, in addition to simultaneous binocular single vision and stereoscopic vision.

Since both the first and the second of these targets depend on the state of development and training of the optomotor reflexes, we may define the sole aim of our treatment as: *the arousing, training and stimulating of the normal optomotor reflexes.*

In this treatment, the logical route to choose is that followed in normal development.

First of all, however, before this constructive treatment can be started, it is necessary to undo everything which has developed abnormally (as compensation reaction) in consequence, of the circumstances obtaining. Thus, any amblyopia, inhibition or abnormal binocular junction must be abolished in the first place.

Since in the building up of the visual faculty protopathic vision precedes epicritic vision, it will be necessary to pay due attention first to primitive vision with the relevant monocular reflexes (abduction, adduction, etc.) and the reflexes to conjugate movements (dextroversion, levoversion etc.), which provide for the subcortical conjunction. The provisional aim of this is the arousing and establishment of peripheral fusion and normal binocular junction. This, thus, involves training in simultaneous binocular vision with the *periphery* of the retina.

Not until training of the retinal periphery has led to the attainment of peripheral fusion and normal junction can one proceed to the training of simultaneous binocular vision with the retinal *centre,* the final aim of which is central fusion. By the exercising of this the consolidation of the normal binocular junction is then accomplished.

Once all this has been achieved, attempts should be made to improve and strenghten the binocular single vision by extending the fusion amplitude, while finally the training of stereoscopic vision may be expected to place the crown on the achievement.

Thus, the usual methods of orthopic training, in which attempts are made from the first to attain simultaneous binocular vision and *central* fusion, must be regarded as fundamentally incorrect, since it conflicts with the natural process of development.

It is obvious that in strabismus convergens the stress must be placed on training of the reflexes to conjugate movements and the abduction reflexes, while in strab. divergens the monocular adduction reflexes in particular will need to be stimulated. It is an important point that the fixation faculty of each eye first must be exercised separately.

The use of atropine drops as a therapeutic aid does not appear to be a logical procedure for very young children. The anatomical differentiation of the macula and the ciliary muscle are such as to render the efficacy doubtful in so far as any effect on perception is concerned. After the age of 6 months the use of such drops might be defensible in connection with the repression of macular functions, which would involve a return to the stage of protopathic vision. This might favour the predominance of peripheral stimuli.

The use of atropine drops in the prevention of amblyopia in cases of strabismus unilateralis in somewhat older children, is a rather different matter. The abolition of amblyopia may indeed promote the achievement of a correct position of the eyes, by banishing the inhibition of the eye not treated with drops, so that a vicious circle is broken. However, if it is also hoped in this way to achieve a parallel position of the eyes by increasing the visual acuity and exercising the central fusion, the procedure would appear to be not altogether theoretically correct, although the effect of such treatment in young children may, for other reasons, turn out better than would be expected on these grounds.

The above considerations apply to the training of reflexes evoked from the peripheral parts of the retina.

Our second aim in treatment, the achievement of a good visual acuity in both eyes, brings us into the domain of the reflexes evoked from central parts of the retina (fusion and re-fixation reflexes). Here we are concerned with the prevention and cure of amblyopia.

Since the best visual acuity will be reached when each retinal element is connected with a sharply dosed optomotor stimulus, the training of visual acuity means the training of the monocular optomotor central reflexes. (See chap. VII).

The simplest way of achieving this is by uninterrupted covering of the good eye for a certain period of time. In this way the amblyopic eye exercises not only its central but also its peripheral reflexes.

It is logical when amblyopia is found to begin with the correction of this fault. Then follow the other exercises, in so far as they are necessary. This order of procedure does not, however, appear to be essential for the achievement of a correct position of the eyes.

It goes without saying that treatment must also aim at the prevention of anything tending to promote the development of a secondary squint. A very important point here is undoubtedly the prescription of the most thorough possible correction of the refraction errors.

Finally, the great value of a rational general treatment must be borne in mind. It is possible that hormonal stimulants will also be found indispensable in the future.

Once the peripheral fusion has been succesfully established, the binocular junction will also be gradually built up at the same time. Once it has been formed, the higher binocular reflexes will develop spontaneously on this basis and it will thus be possible to speak of complete recovery from the affection.

SUMMARY

THE need for further investigation is demonstrated and the lines along which this should proceed are indicated. After a short historical introduction the principal theories on strabismus: the theory of mechanical, peripheral factors; Donder's refraction theory; Worth's fusion theory; Van der Hoeve's theory, Duane's innervation theory and Zeeman's reflex-theory are subjected to critical examination. The great variety of opinions can be accounted for by the fact that the theories lack anatomical and physiological foundations, while the morbid anatomy of squint is still unknown. When, further, we consider our insufficient knowledge of the prenatal development of the brain and of the eye with its central connections — and our still more defective knowledge of their postnatal development extending over several years — it does not appear surprising that now one and now another of the noticeable features of squint was brought forward as the cause of the affection. Peripheral, mechanical causes; abnormal length or weakness of muscles; aplasia or obstetric lesions of the oculomotor nuclei; refraction anomalies (on the grounds of the association between accommodation and convergence); weakness of binocular vision, weak fusion tendency or an excess of inhibition; diplopia-phobia and purposive squint and abnormal stimulation of the convergence centre — all these have had their turn but none of them has survived the serious criticisms accumulated in the course of years.

For instance, in the search for mechanical, peripheral factors, use was made exclusively of optical stimuli for the elicitation of eye movements. If labyrinthine or vestibular stimuli had also been employed, other conclusions would have been reached. The protagonists of the refraction theory were insufficiently aware of the fact that the anatomical and physiological conditions of the child eye at the time of onset of squint, are very far from being comparable with those of the adult eye, on which they based their theories. This error was, in fact, made by investigators in general and it constituted — together with the insufficient knowledge as to the true age of onset of squint — the chief reason why the investigations repeatedly came to a dead end.

A completely new investigation from a new starting point was thus necessary. The first impulse to this was given by Zeeman, as a result of his views — gained in his continuation of the work of Pavlov — on the origin and nature of the optomotor reflexes. Since in the meantime the simple dominant hereditary transmission of squint had been ascertained (Waardenburg) and the rôle of the fusion reflexes (Roelofs) and the significance of amblyopia (Van der Hoeve) had become evident, the lines along which renewed study must proceed seemed to be clearly mapped out.

It was necessary, by means of clinical examination, to acquire a body of data which would suffice, when combined with the existing anatomical knowledge, to build up a theory on a satisfactory anatomical and physiological foundation.

While Zeeman's views had made it clear that an abnormal development of the optomotor reflexes must be given serious consideration in connection with the origin of squint, the question of whether there was an individual or group predisposition to this abnormal development and whether these predisposing and hereditary factors were demonstrable, remained open, as did also that of an explanation for the heterophoria and the squinting position.

In the study of squinting children reported here, the following fixed programme was always used; anamnesis; position of the eyes; ocular movements and optomotor reactions; pupil reaction; refraction; fundus; visual acuity; optical localization; general condition and behaviour were investigated in that order and, finally, due attention was paid to any other abnormalities present. Special emphasis is placed on certain features (anamnesis, optomotor reactions and refraction) of this schema, which is explained in full. The findings in connection with refraction are presented in the form of a graph, showing that nearly 80 % of squinting children under 6 years with hypermetropia have a refraction error in the range 1½ and 4 D, with a peak in the range 2½—3 D. Pronounced errors of refraction and astigmatism are seldom seen in these young children. Emmetropia was not found at all; myopia in only 1.1 % of cases.

On the grounds of comparison with normal children and of

the anatomical condition of the eye at the time of onset of squint, it is concluded that refraction can have no more than a subordinate rôle in the causation of squint, and that in cases starting at a very early age its influence can, a priori, be excluded altogether.

The age of onset was not known with certainty and the available data did not appear to be entirely reliable. In the literature it is assumed that this age lies between the 2nd. and the 4th. year. In a study of 656 cases of squint in children under 7 yr., including 50 under 1, 123 under 2 and 344 between 4 and 6 yr., it was ascertained that in 18.4 % of these cases the squint had appeared very shortly after birth. By the end of the first year 53.9 % had already produced a squint and at the end of the 2nd. yr. as many as 78.3 %. By squint is understood not the dissociated eye movements often seen in young children, but an essential, continuous squint.

From 6 months to 1 year there is a very pronounced decrease in the number of cases.

Between the ages of 18 months and 2 years a marked decrease in the number of new cases is seen; this decrease then proceeds in jumps, so that only an occasional case of a squint beginning after the age of 4 yr. is seen.

These findings are shown on a graph. From the shape of the curve certain deductions as to the origin and course of the affection could be drawn, e.g. that all children are born with the potentiality of strabismus. This view is supported by the other findings. From the study of the age of onset the following conclusions could be drawn:

(1) squint manifests itself at a very early age in the majority of cases;

(2) there are many indications for the probability that all children are born with the potentiality of squint;

(3) correcting and normalizing influences make themselves felt in all children at about the 6th. month;

(4) the resulting process of directing and linking of the two eyes is normally complete at the age of about 18 months;

(5) primary or secondary, endogenous and exogenous factors may disturb this physiological process, with squint as a result.

It appeared thus, that the cause of squint had to be sought in a disturbed course of a physiological process through which all children have to pass in a certain period of their development.

Statistical investigation confirmed anew the markedly hereditary nature of squint; the proportion of hereditary cases among the very young patients amounted to 53 %, while in 44 % of these cases the affection was familial.

The next part of the investigation consisted of a detailed study of the optomotor reflexes under normal and pathological conditions. Optomotor reactions were studied first in normal newborns, then in squinting infants and finally in children with 'papilla grisea' or similar affections. In the last-mentioned group, the children are born blind and the development of their sight and optomotor reflexes can be followed in 'slow-motion', so that they constitute ideal subjects for such an investigation.

Optomotor reactions were first studied in 85 normal newborns. In 2 cases they were tested before light stimuli had had the opportunity of acting. In this way we could get an impression of the eye movements before birth. It was found that in normal neonati the optomotor reflexes to conjugate eye movements predominated and could most readily be elicited from the nasal retinal halves. The monocular reflexes in these children did not appear until much later.

The optomotor reactions of several hundred squinting babies and young children were then tested. The first thing noticed was that the angle of strabismus was continually changing, both with the eyes at rest and upon movement in the horizontal field of gaze. In many cases it was not possible to attain maximum abduction. This dissociation of the eyes was not the result of abducens paralysis, as could easily be demonstrated by means of labyrinthine stimulation.

Still more surprising were the results on stimulation of the nasal and temporal retinal halves of each eye separately in the dark-room. Upon illumination of the temporal half of the retina (uncrossed path) we saw in the overwhelming majority of cases a monocular movement nasal wards, while the unilluminated eye remained at rest. But when the nasal half of the retina (crossed path) was illuminated, no reaction was seen in the majority of cases; only occasionally was there a conjugate movement in the direction of the stimulated eye, but never a monocular abduction movement. From this it is concluded that in the crossed pathways the connection with the binocular gaze centre predominates over that with the monocular, while in the uncrossed pathways there is a predominance of the connection with the monocular over that which the binocular centre.

In contradistinction to normal children, thus, squinting infants show a marked predominance of the monocular adduction reflexes over the reflexes to conjugate movements and the abduction reflexes. The reflexes to conjugate movements do not appear until much later in these cases.

The study of a dozen infants with 'papilla grisea' (Beauvieux' 'dysgénésie myélinique optique des nouveaux-nés'), all of whom in the course of time developed a squint, showed that in such cases — in which it is assumed that the myelination of the optical pathways is disturbed — the first reaction manifested is again a monocular adduction, while here also the reflexes to conjugate movements do not appear until much later.

The development of these children showed many features in common with that of squinting children; in view of this and of the myelination stage of the brain in normal children at the time of birth it seems reasonable to assume that the cause of the disturbances in development of the optomotor reflexes is the same for both groups and consists in the presence of gaps and defects in the myelination of the optical pathways. The degree of disturbance, however, is widely different in the two groups.

By means of 3 diagrams devised by Roelofs, some explanation of the optomotor reflexes is then given and the mode of origin of heterophorias, peripheral fusion and binocular junction is sketched. It is proposed that the term 'retinal correspondence' be replaced by *binocular junction,* so that we may speak of a normal and an abnormal binocular junction.

In addition to the subcortical conjunction, a cortical conjunction will also be established via the binocular junction.

The results of the study of optomotor reflexes are collected into 12 conclusions, the chief of which is that in convergent squint there is a disturbance in the manifestations of the monocular and binocular optomotor reactions. The position of the eyes in strabismus is a direct consequence of this.

By means of electroretinography it could be shown that the cause of this was not in the retina.

In a detailed study of 12 cases observed personally by the writer, the case-histories of which are given, the clinical picture of 'papilla grisea' and similar affections believed to be associated with myelination disturbances in the central nervous system is analysed. Particular attention is paid to the eye movements of these children before and during the development of the optomotor reflexes. These movements were filmed.

The results of investigations on myelination in general and of the optic tracts show it to be probable that myelination disturbances or defects — in a more general sense, as defined [1]) — in every child at the time of birth may account for the observed phenomena. It is suggested that the disturbances observed in the above-mentioned 12 cases should be regarded as expressions of a retardation in the normal growth process of the brain and the name 'myelogenesis retardata' is proposed for this condition. In addition to optical disturbances, this might involve motor and sensory disturbances in other parts of the brain as well, such as were actually observed in some cases.

In view of the analogy in development between the cases of squint and those of myelogenesis retardata, the writer proposes to regard the strabismus cases as mild forms of the syndrome of myelogenesis retardata. The observed differences are then to be regarded as different degrees of retardation. The observed differences in reaction can be explained in terms of the moment at which birth intervenes in a retarded development of the optomotor reactions as a whole.

The following chapter (VII) is concerned with a further study of the average anatomical stage of development reached by the brain and eye at the time of birth. From this, in conjunction with the observed optomotor reactions in newborns, it is attempted to derive an understanding of the visual possibilities in infants. A detailed account is presented of the writer's study of 85 normal newborns in the Groningen University Obstetric Clinic; taken in conjunction with modern knowledge this leads to the conclusions that in the normal baby at the time of birth there are, physiologically, still innumerable gaps and defects in the development of the optical pathways, and that all intermediate stages of optical perception between complete blindness and a fairly advanced development of the visual faculty may occur in the newborn, so that it is advisable to take the 'sight' of infants on the whole with a pinch of salt.

In view of the normal development of the eye and brain, the disturbances of manifestations of optomotor reactions observed in the infants with 'papilla grisea' or strabismus can be satisfactorily accounted for by retardations in the normal process of growth.

[1]) pag. 111

The normal development, in its different forms, thus provides support for the views set out in the chapter on myelogenesis retardata (Chap. VI).

Although not essential to a study of the cause of squint, the subject of amblyopia is treated in some detail in consideration of its rôle in the strabismus problem and the results of the writer's own investigation of the incidence of amblyopia in cases of squint are presented. Regarding the visual acuity as a function of the monocular optomotor reflexes, he sees amblyopia as due to the fact that these reflexes are insufficiently trained or untrained, or to inhibition. In any case, it is always secondary and therefore cannot be taken into consideration as a cause of squint. The way in which amblyopia develops and the duality of the concept are explained. The possibility that amblyopia may be caused by haemorrhages occurring at birth is discussed and dismissed as small. It is considered that 'primary' amblyopia is of rare occurrence. The idea of hereditary amblyopia is refuted, partly on genetic grounds. Attention is drawn to the great possibility of recovery and to the social importance of the effective combating of amblyopia.

In the last chapter (IX), entitled 'New viewpoints and hypotheses on the origin of squint', the author sums up the results of his investigation and uses these for the construction of a new reflex theory of the origin and development of strabismus.

For convenience this chapter is divided into two sections. The first (physiological) part gives the results of clinical, statistical and physiological investigation and deals with the development of the normal optomotor reactions in their proper chronological order and reciprocal relationships.

After a brief introduction on the object and nature of the reflexes, the various reactions and the stimuli which evoke them are reviewed. Their possible mode of origin is also mentioned. The monocular reflexes (adduction, abduction, etc) and the reflexes to conjugate movements, which together are responsible for subcortical conjunction, are discussed.

The reactions aroused by light as the conditioning stimulus will, as conditioned reflexes, be grafted onto these existing reactions, while, in connection with the increasing complex conditions, the number of synaptic junctions is continually on the increase.

The possible mode of origin of the optomotor reflexes is outlined and the various reactions observed in babies are discussed in their normal succession. Here the opportunity is taken of pointing out the difference between so-called and true squinting in infants and to the applicability or non-applicability of the laws of Sherrington and Hering to be baby's eyes.

In the subsequent discussion of the binocular fixation attention is paid to the fusion reflexes, which together with the reflexes to conjugate movements are responsible for this adjustment. They are essential for the attainment of peripheral fusion, which in its turn leads to orthophorization and to the establishment of binocular junction. Once this has been achieved, the eyes can then be moved in a more or less co-ordinated manner via the monocular reflex paths as well, whereby the cortical conjunction is established.

Upon this binocular junction, which is in fact anatomical as well as functional, will now be superposed the higher binocular reflexes; the bilateral adduction will become co-ordinated into a convergence reflex and, with the achievement of central fusion, the possibility of stereoscopic vision will be created.

The second part of this chapter deals with the pathology of squint. The predominance of the adduction reflexes is ascribed to a delayed development of the whole complex of these reflexes, in connection with which it is assumed that these always manifest themselves in a fixed order determined by evolution. Everything depends upon the moment at which birth takes a hand in this chain of events. Since the electroretinogram showed that the cause of this predominance of the adduction reflexes was not in the retina, it must be assumed that the cause is centrally situated. The most probable localization is in a certain part of the occipital lobe. The correlation between function and myelination is also discussed in connection with the development of this part.

Convergent squint is brought about by a predominance of the adduction reflexes over the reflexes to conjugate movements and the abduction reflexes. It is maintained by incompetence of the fusion reflexes to correct the established deviation. As a direct consequence of this, the binocular junction will not be attained at all or only to an unsatisfactory degree, while the same will apply to the reflexes grafted onto this junction. This fact is of the greatest importance in connection with squint. It leads to various consequences, which are discussed in detail with reference to the schemas presented. At the same time the manner in which a divergent squint may come into being is sketched. The consequences of an abnormal junction are also mentioned, for the sake of completeness.

According to the direct cause squint is divided in primary and secundary squint. Primary cases are produced only by the predominance of the adduction reflexes.

The manner in which secondary squint may occur is then discussed; here refraction anomalies may undoubtedly play an important part.

In conclusion a search is made for possible causes of the retardations which are considered to be of such great importance in the origin of primary strabismus. It is suggested that hormonal influences might deserve consideration in this connection.

A few remarks on treatment of the affection, in so far as these follow directly from the results of the investigation, are appended to the last chapter.

In order to facilitate a rapid survey, the reflex theory on the origin and further development of squint, as devised by the writer, is summed up in a number of conclusions.

CONCLUSIONS

Clinical examination and statistical analysis of 984 cases of squint, 514 of which were in children under 2yr. of age, led to the following conclusions:

(1) All children are born with a potentiality to squint and an almost total dissociation of the two eyes.

(2) Congenital squint does not exist; strabismus cannot occur until the light stimulus is able — in connection with the stage of development of the reflex paths — to produce a motor effect.

(3) Strabismus develops at a very early age. The highest frequency of its first appearance is in the 1st. 6 months. In 54 % of cases the condition is evident at the end of the 1st. year and in 78 % at the end of the 2nd. year. After this the frequency drops off rapidly; only occasionally does a case appear after the 6th. year.

(4) Shortly after birth there begins a developmental process which reaches its highest activity at about the end of the first 6 months and is normally complete at the age of about 18 months; this process has as its aim the directing and coupling of the eyes.

(5) The optomotor reflexes which gradually manifest themselves, correlated with the anatomical and functional development of the central nervous system and the eye, prepare the ground for and introduce this corrective and normalizing process.

(6) This directing and coupling process, the orthophorization, reaches its first climax in peripheral fusion and leads to the establishment of the central conjunction based upon the simultaneous stimulation of the homonymous retinal halves of the right and left eye, i.e. to binocular junction.

(7) In this way the dissociation of the two eyes is changed into an association. This binocular junction is the indispensable basis for binocular and stereoscopic vision.

(8) The transitory squint seen in most babies in their first 6 months must be regarded as the expression of a still labile equilibrium under influence of the developing reflexes. In addition, it is possible that subcortical reactions still occur.

(9) During this first stage the position of each eye is determined separately by the combined influence of the reflex forces affecting this eye. If this influence is such that two opposing forces always balance, the result is for each eye a harmonious equilibrium position such that the visual axes of the two eyes at rest will be approximately parallel. If the resultant of the operative forces compels one eye or both to adopt a non-harmonizing position, the visual axes will not be parallel. The result of this will in most cases be a continuous squint. The angle of squint may vary according to the mode of origin.

(10) Under the influence of its postural reflexes each eye thus, will take up a primary equilibrium position, the two visual axes then being approximately parallel or non-parallel. During the process of orthophorization, which is comparable with Straub's emmetropization process, minor deviations will be completely corrected or may remain demonstrable as heterophorias. Major deviations will persist as manifest primary strabismus.

(11) The process of orthophorization, i.e. the building up of binocular fixation reflexes, may be prevented or impeded by secondary, exogenous or endogenous factors, ·with a continuous secondary squint as result. Possible disturbing factors are impairment of general bodily or mental health; functional or anatomical affections of the brain; infections, especially those affecting the central nervous or bronchopneumonial system; anomalies of the eye or its muscles or ligaments; clouding of the media; severe affections of the retina or choroid; refraction anomalies, especially hypermetropia, aniseiconia and deformities of the skull or orbits.

(12) In cases of primary or secondary squint the establishment of the binocular junction has not been possible and the original dissociation of the eyes has been maintained.

(13) The higher binocular reflexes superposed upon the binocular junction have therefore been unable to develop. These will to a certain extent be compensated for by

the monocular reflexes, which are more or less active in the same direction, or by subcortical reflexes to conjugate movements. Hering's law does not hold good in strabismus.

(14) The reflex chain beginning with peripheral fusion reflexes and ending with central fixation reflexes is also unable to develop.

(15) As demonstrated clinically, the incorrect position of the eyes in strab. convergens is a consequence of the predominance of the monocular adduction reflexes over those for conjugate movements and for abduction. The electroretinogram shows that the cause does not lie in the retina. In all probability this is centrally situated, in the occipital lobe. Strab. divergens develops in consequence of a predominance of the reflexes for abduction over those for adduction, with a continual predominance of the reflexes for conjugate movements.

(16) The disturbance of normal development of the optomotor reflexes may be a manifestation of delayed myelination (myelogenesis retardata).

(17) The clinical type of the strabismus (alternans, unilateralis) is determined chiefly by the presence or absence of a preference for one or the other eye.

(18) Amblyopia and abnormal binocular junction are both consequences of squint and not possible causes of it.

(19) In accordance with its nature (as a growth disturbance), squint shows a marked tendency towards recovery. The number of cases achieving a delayed spontaneous recovery (with respect to the position of the eyes) in the later years of childhood is quite considerable.

(20) The presence of amblyopia in one eye does not prevent spontaneous recovery, as far as the correction of the position of the eyes is concerned.

(21) The correlation between squint and hypermetropia can be explained by the fact that the predominance of the monocular adduction reflexes interferes with an exact training of the central, monocular fixation reflexes.
In its turn the consequently reduced visual acuity hinders the process of emmetropization.

(22) Treatment must be in accordance with the normal process of development and must be started as early as possible.

(23) With a timely start and carefully supervised training of the optomotor reflexes it is, in our opinion, by no means impossible that in future operative treatment will be considered only in exceptional cases.

(24) It is conceivable that hormonal factors are involved in the causation of the retardations which are taken as the basic cause of squint.

BIBLIOGRAPHY.

Abraham. Am. J. Ophth. 32 93, 1949. Am. J. Ophth. 26, 271, 1943.
Arch. o. Ophth. 12, 391, 1934.

Adler. Arch. of Ophth. 33, 362, 1945, cited by Scobee. Quart. Rev.
Ophth. 4 no. 1 March: 10, 1948.

Aegina, Paullus van. 668—685 at Alexandrië, cited by Zeeman Strab.
Symposium 1943, Amsterdam.

Amsler. Ophthalmologica, 110, 225, 1945.

Anderson Ringland. Ocular vertical deviations. Suppl. Br. J. Ophth. 12,
1947.

Ariëns Kappers C. U. The evolution of the nervous system in invertebrates,
vertebrates and man. Haarlem 1929.

Arlt. Die Krankheiten des Auges, Prag 1856.

Bárány. Act. Oto-laryng, VII, 97, cited by Thornval. Act. Oto-laryng VII,
45, 1920/21.

Barie. T.O.S. 43, 612, 1923.

Barkan, O. & H. Arch. o. Ophth. 62, 691, 1931.

Bartels. Uber Augenbew. bei Neugeborenen. Deutsch. Med. Wochenschr.
38, V, 1932.

Bartels & Zika. Graefe's Arch. 76, 93, 1910.

Bartisch. Ophthalmodonleia cited by Duke-Elder. Textb. Ophth. IV, 3810,
1949.

Bauer und Bode. Handb. d. Erbbiol. III, 1940.

Beauvieux. Revue Neurologique 1921. La Pseudo-Atrophie optique des
Nouveau-nés, Paris 1926. Arch. d'Opht. 7, 241, 1947.

Berger. Arch. f. Psychiatr. 33, 1909.

Bergeron. Les manifestations motrices spontanées chez l'enfant, Paris 1947.
Annales médico-psychologiques 1938. La Médicine Infantile 1938.

Best. Cited by Winkelman. Thesis, A'dam 1949.

Bielschowsky. Heidelb. II, XI, 1932. Am. J. 20, 478, 1937. Kl. Monatsbl.
Augenhk. 27, 302, 1926. Br. J. Ophth. 9, 107, 1938.

Bing. Arch. o. Ophth. 20, 175, 1938.

Binkhorst. Toxoplasmosis. Leiden 1949.

Blet. Bull. et Mém. Soc. Franç. d'Opht. 1950.

Bolk. Ned. Tijdschr. v. Geneesk. p. 156, 1922. Ned. Tijdschr. v. Geneesk.
70, 1718, 1926. Ned. Tijdschr. v. Geneesk. 71, 2216, 1927.

Braak Ter & Rademaker. Brain 71, 48, 1948.

Braendstrup. Act. Ophth. 5, 22, 1944.

Braun-Vallon. Ann. d'Oc. 181, 321, 1948.

Brouwer. Cited by Magnus. Körperstellung. 1924. Zeitschr. f. Neur. Psych. XI, 152, 1918. Personal communication 1949.

Burrian & Wald. Am. J. Ophth. 27, 950, 1944.

Cajal, Ramony. Le Chiasme et les Entre-croisements sensori-moteurs, 1898.

Camper. De Oculorum Fabrica et Morbis. Opuscula Selecta. A'dam 1913.

Cat, Le. Cited by P. Camper. De Oculorum Fabrica et Morbis. Opuscula Selecta. A'dam 1913.

Ceasar. Samml. zw. Abhandl. 8, 8, 1912.

Chavasse. Worth's Squint 1939: 7th. Ed. Philadelphia. Cited by Scobee; The Oculorotary muscles. St. Louis 1947.

Clark. J. Anat. XXV, 419, 1941.

Clausen. Kl. Monatsbl. Augenhk. 68, 850, 1922.

Clausen & Bauer. Z. Augenhk. 50, 313, 1923.

Cohn. 1904. Berl. Kl. Wochenschr. XL, 1047, 1904.

Collin, A. Le développement de l'enfant. Paris 1914.

Conel Le Roy. The postnatal development of the human cerebral cortex. Vol. I, II, III, 1939/1947.

Cords. Kurzes Handb. Ophth. III, 491, 1930. Kurzes Handb. Ophth. III, 537, 1930.

Cramer. Anat. H. 10, 1898.

Cridland. Br. J. Ophth. 25, 141, 1941.

Crinis, M. de. Wiener Kl. Woch. schr. II, 1932. Monograph. a. d. Gesammt. Geb. der Neurol. und Psych. 64, 1938

Mc. Culloch and Jefferson. Brit. Med. Bull. 6, 4, 1950.

Czellitzer. Arch. Rassenbiol. 14, 337, 1923.

Déjerine. Anat. du système nerveux I, 1895.

Deloge, Am. J. Ophth. 4, 407, 1921.

Delord. Arch. d'Opht. XXXVIII, 597, 1921. Clin. Opht. 344, 1921.

Descartes. Cited from Serrurier. Leer en leven. 's Gravenhage 1930.

Dieffenbach. Med. Zeitschr. Ver. Heilk. Preussen 46, 1839.

Donders. Ametropie en haar gevolgen. 46, 1860. Anomalies of accommodation and refraction 1864. Cited by v. Hess. Graefe-Saemisch Hb. 2e Aufl. Kap. XII, 460, 1910. Arch. f. Ophth. 9, 1, 1863.

Downing. Am. J. 33, 137, 1945.

Duane. Arch. o. Ophth. 28, 261, 1899. Motor anomalies of the eyes 1897. Contr. to Ophth. Sience 34, 1926.

Duke-Elder. Textb. o. Ophth. IV, 1949.

Duyse, van. T.O.S. LIII, 29, 1933.

Economo und Koskinas. Die cyto-architectonik der Hirnrinde des erwachsenen Menschen. Wien 1925.

Edgerton. Am. Arch. Ophth. 11, 838, 1934.

Engelmann. Cited by Von Hess, Graefe-Saemisch Handb. 2e Aufl. Kap. XII, 460, 1910.

Evans. Am. J. Ophth. 12, 194, 1929.

Feldman and Taylor. Arch. o. Ophth. 27, 851, 1942.

Flechsig, P. Anatomie des Menschlichen Gehirns und Rückenmarks auf Myelogenetischer Grundlage. Bd. I, Leipzig 1920.

Franceschetti. XI Congr. franç. O.N.O. Bordeaux 3. 5. 6. 1939. Rev. O.N.O. 17, 244, 1939. Ophthalmologica 114, 332, 1947.

Fuchs and Pötzl. Jhrb. f. Psych. XXXVIII, 15, 1917. Zeitschr. f. Neurol. 43, 276, 1918.

Garcin & Rademaker. L'Encéphale 1, 17, 1934.

Graefe, Von. Arch. f. Ophth. 31, 177, 1857.

Graefe, A. Graefe-Saemisch Handb. 2e aufl. VIII, 1898.

Gourevitch. Rapport du XIe Congrès de Psychologie. Paris 1937.

Granström. Acta Ophth. 14, 72, 1936.

Granström et Magnusson. Br. J. Ophth. 5, 34, 1950. Arch. franç. de Pédiatrie 1949.

Hagedoorn. Suppl. Symposium Strabismus. Amsterdam 1944.

Haitz. Kl. Monatsbl. Augenhk. 99, 761, 1937.

Halbertsma. Arch. d'Opht. 1937.

Harms. Arch. f. Ophth. 138, 149—148, 1938/39.

Hartmann. Ann. d'Oc. 181, 449, 1948.

Hartridge. Recent advances in the physiology of vision. 1950. Londen.

Hasner, von. Beiträge zur Physiologie und Path. des Auges. Prag 1873.

Heine. Kl. Monatsbl. Augenhk. 42, 10, 1905.

Heinonen. Schweiz. Med. Woch. 44, 1131, 1939.

Henkes. Inleiding tot de clinische electroretinography. Ned. Tijdschr. v. Geneesk. 1949.

Hering. Die Lehre v. binokul. Sehen. Leipzig 1868. Handb. d. Psychologie III, 504, 1879. Cited by V. Hess. Graefe-Saemisch Handb. Kap. XII, 460, 1910. from: Beitrage zur Physiologie V.

Hippel, Von. Arch. f. Ophth. 45, 2, 286.

Hire, La. Cited by Duke-Elder. Textb. o. Ophth. IV, 3809.

Hoeve, v. d. Kl. Monatsbl. Augenhk. 69, 620, 1922. Graefe-Saemisch Handb. 3e Aufl. Operationslehre 1922. Strabismus Symposium. A'dam 1943/44.
Ann. d'Oc. 1, 1917. Ned. Tijdschr. v. Geneesk. 1918, p. 790.

Hoeve, v. d. & Kleyn de. Pflüger's Arch. 169, 241, 1917.

Holmes. Brit. Med. J. II, 107, 1938. Introd. to clin. Neurology. Edinb. 1946. Br. J. Ophth. 449, 506, 1918. T.O.S. I. 253, 1930.

Irvin Rodman. Am. J. Ophth. 27, 740, 1944.

Jackson Hughlings. Cited by Bergeron. Les manifestations motrices spontanées chez l'enfant. 1947.

Jaeger. Ueber die Einstellung des dioptrischen Apparatus 1861.

Jaensch. Kl. Monatsbl. Augenhk. Beiheft 1938.

Juler. T.O.S. 41, 129, 1921.

Kappers Ariëns C. U. The evolution of the nervous system in invertebrates, vertebrates and man. Haarlem 1929.

Karelitz and Vogel. Am. J. Childr. 50 873, 1935.

Karpe. The basis of clinical electroretinography. Act. Ophth. Suppl. XXIV. 1945.

Keith Lyle. Worth's and Chavasse's Squint. London 1950.

Kestenbaum. Arch. o. Ophth. 74, 113, 1930. Am. J. Ophth. 31, 94, 1948.

Kleist. Klin. Woch. schr. I, 1926.

Klüver. J. Psych. 11, 23, 1941. A. of neurol. Psych. 42, 979, 1939. Biological Symposia 7, 253, 1942.

Kurz. Okulo digitálin reflex. Ceskoslov. Ophthalmologie. Prague 1949.

Lachman. Am. J. Ophth. 10, 164, 1927.

Lagleyze. Du strabisme. Paris 1913.

Lagrange & Moreau. Arch. d'O. 27, 209, 1907.

Lancaster. Am. J. Ophth. 24, 5, 1941.

Lesné et Richet. (fils) Presse Médicale I, 18, 1925.

Levinson and Stillerman. E.E.G. Journal, May 1950.

Magitot. Ann. d'Oc. 1909.

Magnus. Körperstellung. Berlin. 1924.

Mann, Ida. Developmental abnormalities of the eye. 23, 1927, Cambridge. The developing third nerve in human embryos. J. of Anat. LXI, IV, 1927. The development of the human eye. Cambridge 1928.

Marquis. Brain 60, 1, 1937.

Meller. Wiener Kl. W. Schr. 39, 190, 1926. Arch. f. Ophth. 33, 289, 1947.

Middlemore, R. Cited by Campbell. Br. J. Ophth 31, 332, 1947.

Minkowski. Schweiz. Arch. Neurol. 6, 1920. Revue Neurol. 1921. Schweiz. Med. Woch. schr. 754, 1922. L'état actuel des réflexes. Paris 1927.
T. VIII de l'Encyclopédie Franç. p. 10 Paris 1938.

Miranda. A. d. Oft. H. A. 35, 393, 1935.

Monakow, von. Cited by Bergeron. Les manifestations motrices spontanées chez. l'enfant. Paris 1947. p. 92.

Nordenson. Acta Chir. Scand. 52, 45, 1919/20.

Nordlöw. Acta Ophth. 1942. Suppl. XX.

Offret, d'. Soc. d'Opht. de Paris 1942.

Onfray, R. Bull. et Mém. de la Soc. Franç. d'Opht. 25, 1949. Traité d'Ophtalm. VII. p. 31—136, 1939.

Panum. Physiologische Untersuchungen über das Sehen mit zwei Augen. 1858.

Parinaud. Rapport sur le traitement du strabisme. Soc. Franç. d'Opht. Paris 1893.

Parsons. Introduction to the Theory of Perception. Cambridge 1927.

Pavlov. Conditioned Relfexes. Oxford 1927.

Peter. Am. J. O. 15, 493, 1932 and 16, 481, 1933. The Extra ocular Muscles. Philadelphia 1941.

Pfeiffer. Monogr. a. d. Gesammt Gebiete d. Neur. und Psych. 43, 1925.

Piper. Arch. f. Ophth.148, 555, 1948.

Poppelreuter. Zeitschr. f. Ges. Neurol. Psych. 83, 26, 1923.

Poulard. Ann. d'Oc. 158, 95, 1921.

Priesthly Smith. Trans Ophth. Soc. U. K. 18, 17, 1898. Int. Congress Ophth. Utrecht 1899.

Pugh. Br. J. Ophth. 18, 446, 1934.

Rademaker et Garcin. L'Encéphale 1, 17, 1934.

Rademaker & Ter Braak. Brain 71, 48, 1948.

Reddingius. Zeitschr. f. Psych. und Physiol. der Sinnesorgane. Bd. 21, 420, 1899.

Rhenter. Pratique de l'Art des Accouchements. Paris 1928, cited by Bergeron.

Rieken. Cited by Winkelman. Thesis A'dam 1949. Arch. f. O. 145, 1943.

Rochat. Oogheelk. v. d. Algem. Arts. 1934.

Rochon-Duvigneaud. Bull. Soc. Franç. d'Opht. 1, 1933.

Roelofs C. Otto. Arch. f. Ophth. 85, 66, 1913. Arch. f. Ophth. 113, 239, 1924. Arch. f. Augenhk. 97, 229, 1926.

Roelofs C. Otto & v. d. Waals. Zeitschr. f. Psychol. 13, 5, 1935.

Roelofs C. Otto & Zeeman. Arch. f. Ophth. 88, 1, 1914. Arch. f. Augenhk. 98, 238, 1927.

Roesch. Actualités scientifiques 954, Paris 1943, cited by Zeeman. Ophthalmologica 118, 254, 1949.

Roux. Arch. de Neurologie VIII, 177, 1899, cited by Muskens. Das Supra-Vestibuläre System. Amsterdam 1934.

Le Roy Conel. The Postnatal Development of the human cerebral cortex. Vol. I. The cortex of the new-born 1939. Vol. II. The cortex of the one-month infant 1941. Vol. III. The cortex of the three-month infant 1947 London.

Rubino et Pereyra. Bull. et Mém. Soc. Franç. d'Opht. 1950.

Sattler. Zeitschr. f. Augenhk. 63, 19, 1927.

Scherb. Festschrift Vogt. 1939.

Schiötz. Acta Ophth. 5, 285, 1927.

Schweigger. Kl. Untersuchungen über das Schielen. Berlin 1881.

Schweigger und V. Graefe. Spec. Aug. 124, 1873.

Scobee. The oculorotary muscles. St. Louis 1947. Quarterly Rev. Ophth. 4/1, 10, 1948. Am. J. O. 31, 794, 1948.

Sherrington. The integrative action of the nervous system, 1947.

Sicherer, Von. Münch. Med. Woch. IV, 4029, 1907.

Sicherer, Von und Stumpf, M. Geburtsh. und Gynäk. 13, 408, 1909.

Smukler. Am. J. Ophth. 16, 621, 1933.

Sorsby. Cited by Hartridge. Recent advances of the physiology of vision. London 1950.

Stenström. Cited by Hartridge. Recent advances of the physiology of vision. London 1950.

Stenvers. Cited by Magnus. Körperstellung 1924. Berlin. p. 123, 126, 140.

Straub. Arch. f. Ophth. p. 70—130 (1909). Ned. Tijdschr. v. Geneesk. p. 1970—1980 (1910). Geneesk. Bladen p. 105—150 (1915).

Stutterheim. Squint and convergence. London 1946. British J. Ophth. Suppl. V. 1931.

Swan. Arch. o. Ophth. 37, 444, 1947.

Symposium Strabismus. Neth. Ophth. Soc. Amsterdam 1943/44.

Thums. Monogr. a. d. ges. Geb. d. Neur. und Psych. 66, 199, 1939.

Travers. Br. J. Ophth. Suppl. 1936. Br. J. Ophth. 21, 58, 1940.

Uhthoff. Kl. Monatsbl. Augenhk. 78, 453, 1927.

Verhoeff. Arch. o. Ophth. 19, 663, 1938.

Viallefont. Journ. d'Opht. du Midi 1943.

Virchow. Cited by Bergeron. Les manifestations motrices spontanées chez l'enfant Paris 1947. p. 95.

Vogt. Kl. Monatsbl. Augenhk. 103, 291, 1939.

Volkmann. Zeitschr. Anat. 85. H. 5/6, 561, cited by M. de Crinis. Monogr. Gesammt. Geb. d. Neurol. und Psych. 64, 1938.

Waardenburg. Ned. Tijdschr. Geneesk. II 47, 5783, 1928. Symposium Strabismus. Amsterdam 1943.

Wald and Burrian. Am. J. Ophth. 27, 950, 1944.

Wecker, De. Cited by Parinaud. Rapport sur le traitement du strabisme. Soc. Franç. d'Opht. 10, 1893.

Weil. Influence of sex hormones upon the chemical growth of the brain of white rats. Growth VIII, 2, 1944.

Wernicke. Arch. f. Psych. 20, 272, 1889.

Westphal. Arch. f. Psych. 29, 1897.

Wirth. Zeitschr. f. Augenhk. 26, 326, 1911.

Worth. Squint. 6 th. Edition 1929.

Zamenhof. Stimulation of the proliferation of neurons by the Growth Hormones. Growth V, 2, 1941.

Zeeman. Arch. f. Ophth. Bd. 78. 1911. Arch. f. Augenhk. 100/101. 1. 1929. Over twee-oogig zien. Rede Amsterdam 1936. Strabismus Symposium. Neth. Ophth. Soc. Amsterdam 1943/44. Ophthalmologica 118, 254, 1949.

Zeeman & Roelofs C. Otto. Arch. f. Ophth. 88, 1, 1914.

CONTENTS.

CHAPTER I

CHAPTER II

CHAPTER III

CHAPTER IV

CHAPTER V

CHAPTER VI

CHAPTER VII

222